D0989611

BIOLOGICAL PHYSICS SERIES

BIOLOGICAL PHYSICS SERIES

Continued after index

M. Zamir

The Physics of Pulsatile Flow

Foreword by Erik L. Ritman

With 75 Illustrations

M. Zamir
Department of Applied Mathematics
University of Western Ontario
London, Ontario N6A 5B7
Canada
zamir@julian.uwo.ca

Cover illustration: Pulsatile flow in a rigid tube compared with that in an elastic tube, with a vascular tree structure in the background.

Library of Congress Cataloging-in-Publication Data
Zamir, M. (Mair)
The physics of pulsatile flow/M. Zamir.
p. cm. — (Biological physics series)
Includes bibliographical references and index.
ISBN 0-387-98925-0 (hardcover: alk. paper)
1. Blood flow. 2. Tubes—Fluid dynamics. 3. Body fluids.
I. Title. II. Series.
QP105.4.Z35 2000
612.1'181—dc21 99-42457

Printed on acid-free paper.

Production managed by Frank McGuckin; manufacturing supervised by Jeffrey Taub.
Digitally produced from the author's LaTeX files using Springer's svsing.sty macro.
Printed and bound by Edwards Brothers, Inc., Ann Arbor, MI.
Printed in the United States of America.

9 8 7 6 5 4 3 2 1

ISBN 0-387-98925-0 Springer-Verlag New York Berlin Heidelberg SPIN 10743202

To Lilian, Jonathan, Marta, Emily, and David,
for giving five clear meanings to my life.

Foreword

Classic texts in the field of analysis of flow in blood vessels have been written over the years and what these say is still valid today. However, our knowledge of pathophysiological mechanisms has changed with increasing rapidity over the past 20 years, as has our ability to visualize the three-dimensional geometry of blood flow and blood flow velocity distribution within the *in vivo* blood vessels. Consequently, with the increased need to fully exploit the new imaging capabilities and our additional biological knowledge, this book is a welcome addition to our armamentarium used to achieve those new goals.

Whereas in the past pulsatile flow (and consequent wave reflections) was often seen as "frosting on the cake" of analysis of blood flow problems or perhaps as an issue that should be understood only in a general sense, our new capabilities and understanding require more accurate analyses of specific systems, not just of constructs based on statistical data describing a vascular tree. Examples of this new need include the situation where the detailed branching geometry of an arterial tree is known from imaging and it is desired to see to what extent local fluid dynamic characteristics can explain the specific localization of disease such as atherosclerosis, or of the extent to which the heterogeneity of perfusion throughout an organ can be attributed to the vascular tree branching geometry or to the mechanical properties of the vascular walls. Another area that is now opening up rapidly involves the design of hybrid synthetic scaffolds of blood vessels with cellular (especially endothelial cells) lining of the scaffold. The local shear stress distribution on the scaffold and the cells attached to it must now be examined specifically for these prostheses.

Although imaging methods such as Doppler ultrasound and magnetic resonance imaging can provide direct indices of local blood flow velocity distributions, the spatial resolution is limited so that the use of analysis such as described in this book along with the detailed three-dimensional geometry of the vascular tree and its wall, will provide the high-resolution distribution of shear stress and local pressures needed for detailed analyses. Such analyses should permit the assessment of the relative importance of those variables in, for instance, the stimulation or suppression of the genetic expression of the homeostatic responses in such locations.

This book is not a list of recipes of analytical methods. The incorporation of basic physics and mathematical derivation provide the background understanding necessary for the serious student to evaluate the value of this approach in these types of applications. Consequently, the expected accuracy and contribution of pulsatile flow to the pathophysiology of vascular remodeling and acute response to changes in the hemodynamics can be evaluated with greater confidence.

In conclusion, this book is relevant and timely in today's rapidly developing field of pathophysiology of the vascular tree and the ready availability of powerful imaging and computational facilities.

Rochester, MN

Erik L. Ritman, M.D., Ph.D.
Professor, Physiology and Medicine
Ralph B. and Ruth K. Abrams Professor
Mayo Clinic Graduate School of Medicine

Series Preface

The field of biological physics is a broad, multidisciplinary, and dynamic one, touching on many areas of research in physics, biology, chemistry and medicine. New findings are published in a large number of publications within these disciplines, making it difficult for students and scientists working in biological physics to keep up with advances occurring in disciplines other than their own. The Biological Physics Series is intended therefore to be a comprehensive one covering a broad range of topics important to the study of biological physics. Its goal is to provide scientists and engineers with text books, monographs and reference books to address the growing need for information.

Books in the Series will emphasize frontier areas of science including molecular, membrane, and mathematical biophysics; photosynthetic energy harvesting and conversion; information processing; physical principles of genetics; sensory communications; automata networks, neural networks, and cellular automata. Equally important will be coverage of current and potential applied aspects of biological physics such as biomolecular electronic components and devices, biosensors, medicine, imaging, physical principles of renewable energy production, and environmental control and engineering.

We are fortunate to have a distinguished roster of consulting editors on the Editorial Board, reflecting the breadth of biological physics. We believe that the Biological Physics Series can help advance the knowledge in the field by providing a home for publications in the field and that scientists and practitioners from many disciplines will find much to learn from the upcoming volumes.

Oak Ridge, Tennessee

Elias Greenbaum
Series Editor-in-Chief

Preface

The core material of this book is based on lectures given by the author over a period of 30 years to graduate and undergraduate students in science, engineering, and medicine. The topics are extracted from courses in fluid dynamics and blood flow and are here pieced together to form a self-contained course unit on pulsatile flow. The material is particularly suitable for a senior undergraduate or graduate course, as it contains all the required background in fluid dynamics, physics, and mathematics.

The book is divided into six chapters. The first chapter deals with preliminary concepts from fluid dynamics, continuum mechanics, and blood flow. The second chapter contains a fully detailed derivation of the Navier-Stokes equations and equation of continuity. The third chapter deals with steady flow in a tube and its application to vascular tree structures. The fourth chapter deals with pulsatile flow in a rigid tube, with all the elements of the classical solution of this problem. The fifth chapter deals in a similar way with the more complicated problem of pulsatile flow in an elastic tube. The sixth and final chapter deals with the phenomenon of wave reflections, its effect on flow in tubes and in vascular tree structures, and the specialized equations required to deal with the phenomenon. Smaller sections in each chapter deal with smaller subtopics and are designed to provide, approximately, the material for one lecture.

Pulsatile flow is generally studied in association with blood flow, and blood flow in turn is generally studied in association with cardiovascular function and disease. Because of these associations, books on the subject have tended to present pulsatile flow only within the larger context of the anatomy, physiology, and pathology of the cardiovascular system and of clinical aspects of cardiovascular disease. The two classical books of McDonald and of Milnor have carried and continue to carry this mission dutifully and with invaluable service to the scientific community. References to these and to a number of other books on the subject appear at different points in the text.

In this book, by contrast, and as the title indicates, pulsatile flow is presented in the context of physics and, by implication, mathematics. The motive for this is that pulsatile flow *is* a physical phenomenon, and its description and understanding involve a considerable amount of mathematical analysis and results. An understanding of these results requires a good grounding in the physics of fluid flow in order to understand the basis of the equations on which the results are based. They also require full details of the solution of these equations, including all the assumptions on which the solutions are based, because only then is it possible to interpret the results correctly and to best advantage.

This "indulgence" in mathematics and physics has not been afforded to pulsatile flow in the past, because, as stated above, the subject has usually been presented in the context of cardiovascular function and medicine. The amount of analytical detail involved in this subject would require a prohibitive amount of space in that context, therefore a reference would normally be given to published papers instead. The required details can indeed be obtained from classical papers in which these solutions were first presented (references in the text), but the exercise requires proficiency in the physics of fluid flow and in the mathematical solution of differential equations, which again must be obtained from other sources. The impetus for the present book was the challenge of bringing all these elements under one cover and dealing with them in sufficient detail as to, in effect, bring the subject of pulsatile flow to the level of a textbook.

The focus on physics and mathematics in this book is not intended to detract from the ultimate association of pulsatile flow with cardiovascular function and disease. On the contrary, it is the author's belief that the best way to deal with that association is to give those involved in work or study of cardiovascular function and disease the mathematical and physical tools required for dealing with the subject. To make these tools available, it is necessary not only to present the results of the classical solutions of pulsatile flow problems but also to present the physics and mathematics that give these solutions their meaning and legitimacy. It is necessary not only to recite and abide by all the assumptions involved but also to see how and precisely where these assumptions arise in the analytical process, so as to assess the degree of their relevance in each specific problem.

This book is intended to make these tools available, with the hope of making the subject of pulsatile flow more accessible to those who are not proficient in every element of the subject. The rationale for this approach is that applications of this subject to cardiovascular function and disease, whether by a medical researcher or a biomedical student, will be better founded and produce more fruitful results when they are based on a good understanding of the physics and mathematics involved.

The questions of whether or not blood is a Newtonian fluid, and whether or not wave reflections are important in the cardiovascular system, are two good examples in which mastery of the mathematical and physical tools is important in order to deal with these issues in individual situations. Neither question has a "yes/no" answer that is valid under all circumstances. Both are important questions that must be considered in each case and in the context of mathematics and physics. It is insufficient to refer to the corpuscular structure of blood and conclude that blood is not a Newtonian fluid; the question must be examined in each case with a clear understanding of the continuum concept on which the equations of fluid dynamics are based. It is insufficient to observe the adverse effects of wave reflections in one case and conclude that the effects will be adverse in all cases. The effects must be calculated individually in each case, and to do so requires the appropriate mathematical and physical tools. In fact, wave reflections are highly widespread in the cardiovascular system because each one of the billions of vascular junctions acts as a potential reflection site. On purely biological grounds, therefore, it is unreasonable to surmise that the functional design of the cardiovascular system is such that wave reflections produce adverse effects throughout the system.

I owe my training in fluid dynamics to two teachers at the University of London whose influence on me has been so deeply ingrained that I can barely trust a thought on the subject as being entirely my own. Professor A.D. Young, FRS, who taught me to understand boundary layer flow as it should be understood, and Professor R.D. Milne who taught me the power of mathematics in fluid dynamics. I owe my excursions into blood flow to several colleagues whose influence on me was pivotal: Professor Margot R. Roach of the University of Western Ontario who practiced cardiovascular medicine with an open door into physics, and the late Professor Alan C. Burton who ensured that the door remained open. Together they were responsible for my entry into the field. Professors M.D. Silver of the University of Toronto and G. Baroldi of the University of Milan who taught me how far the subject of fluid dynamics is from that of clinical pathology, and how hard, yet rewarding, it is to bring the two subjects together in the cardiovascular system. Professor R.E. Mates of the State University of New York at Buffalo whose work on pulsatile flow in the coronary circulation has been a constant inspiration to me, and Professor E.L. Ritman of the Mayo Clinic at Rochester, Minnesota, whose work on the visualization of blood vessels has been of particular interest and value to me. Professor G.A. Klassen of Dalhousie University who has given me so generously of his expertise in cardiology and in the physics of the cardiovascular system. Our numerous discussions over the years have made a significant contribution to my education in the field. Finally, the interdisciplinary environment essential for work on this subject was provided to me by two very special departments here at the University of Western Ontario; Applied Mathematics and Medical Biophysics. I am indebted to my colleagues in both departments.

I owe a special debt of gratitude to my friend and colleague Professor S. Camiletti who so kindly undertook the tedious task of reading the manuscript before going into print. I have often turned to his wonderful command of mathematics and fluid dynamics to test or clarify this or that idea, and rarely left the discussion disappointed. Getting the manuscript into his hands was my first and almost instinctive thought, and I am deeply grateful to him for accepting the task. I am also indebted to Hope Woolley for proofreading the problems and answers sections, with meticulous care.

I owe thanks to my long time aide and dearest friend, Mrs. Mira Rasche, who walked the road from physics to medicine with me and made the journey so much easier. Special thanks are also due to Ian Craig for skilful photography work over the years, especially with changing technology. I am grateful to Dr. Elias Greenbaum, Editor-in-Chief of the Biological Physics Series, and to Maria Taylor, its Executive Editor at Springer-Verlag, for granting me an entry into this exciting venture. Their enthusiasm for the series and for the book has been a sustaining help to me. Many thanks are also due to Frank McGuckin, Production Editor, for orchestrating the final stages of the book with admirable patience.

London, Ontario, Canada M. ZAMIR

Contents

1 Preliminary Concepts

1.1 Flow in a Tube

Flow in a tube is the most common fluid dynamic phenomenon in biology. For two good reasons, the bodies of all living things, from the primitive to the complex, plant or animal, are permeated with a plethora of fluid-filled tubes (Fig.1.1.1). Fluids provide an indispensable medium for convection and diffusion of the numerous chemical and biochemical products required for the maintenance of biological function, and tubes provide the most efficient medium for the containment and transport of these fluids.

The evolution of flow in tubes as a tool in biology is so closely intertwined with the evolution of living organisms that it is difficult to view the two as separate from each other [1–4]. Not even in the realm of science fiction is it possible to imagine the evolution of a living organism of any degree of complexity without the facility of flow in tubes. Only the most primitive single-cell objects manage to function without this facility, although, even here, fluids are used to facilitate transport within the cell and to enhance chemical and biochemical reactions [5,6].

There is likely more fluid-filled tubes in a single human body than in any other body or place on this planet. This can be observed both on the scale of phylogenic development of species as on the ontogenic scale of embryonic development. Organisms of any degree of complexity require the facility of flow in tubes, more so as the degree of complexity increases. The development of the vascular system in the human embryo, which has been well described in the literature, illustrates well how the emergence of

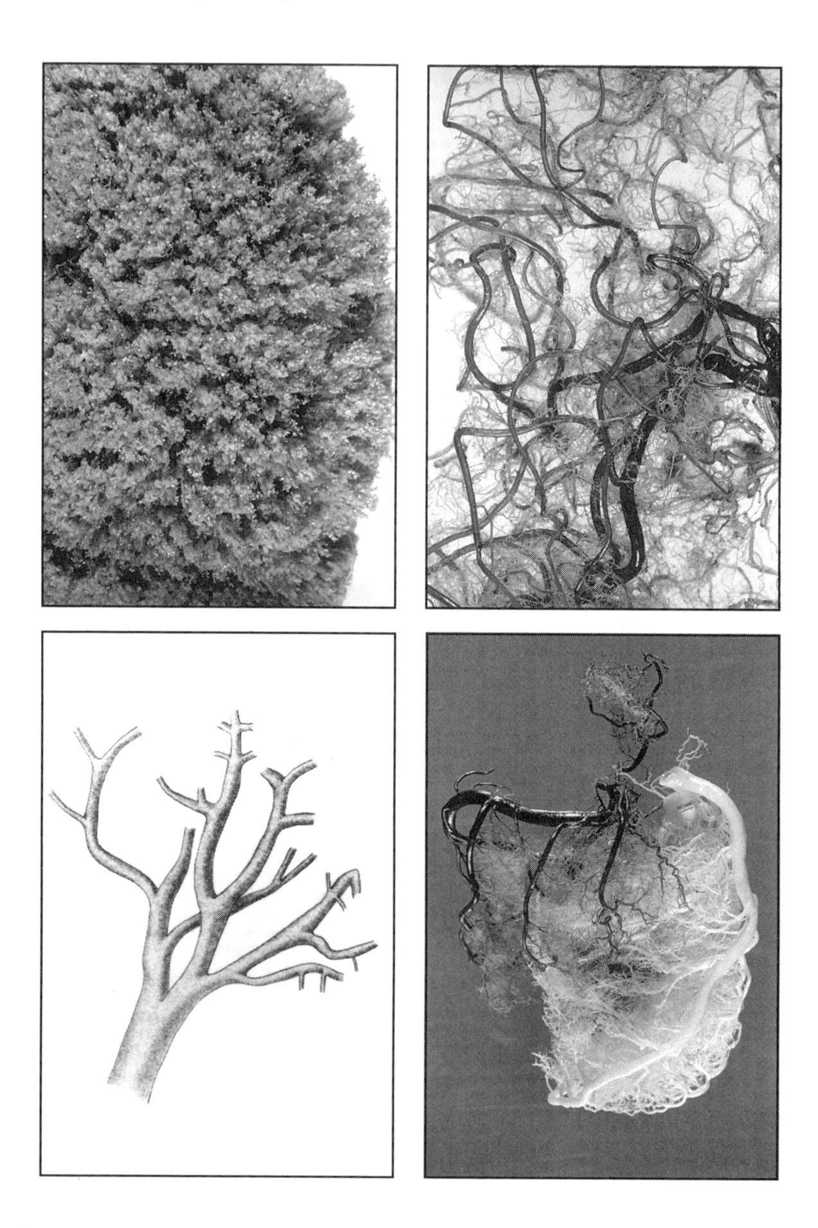

FIGURE 1.1.1. Flow in tubes is an inseparable part of biology. Clockwise from top left: vasculature of the kidney, brain, heart, and a close up of underlying tree structure.

the tubular structures of the vascular system is an inseparable part of the growth of the embryo itself [7,8].

1.2 What Is a Fluid?

As a noun or adjective, the term "fluid" refers to a *state* of matter, not a substance. In elementary physics we learn that matter exists in three distinct states: solid, liquid, and gas. The term "fluid" refers to the second and third of these states collectively. A fluid is a liquid or a gas, *any* liquid or gas. This word usage is not only justified but appropriate when studying the dynamics of fluids, because these dynamics depend on the *mechanical properties* of fluids, not on their chemical substance. What are the mechanical properties that identify a fluid body?

By a somewhat intuitive sense we find it easy to identify a body of water as a fluid body and a body of steel as a non-fluid body, but we find it less easy to identify the properties that make the distinction between the two. Often the word "flow" is used to explain that water can flow and steel cannot, but this compounds rather than resolves the issue, for the word "flow" is itself in want of definition. Only careful mechanical considerations can resolve the issue. The mechanical properties of a material body have to do with the way in which the body responds to an attempt to deform it [9–11]. We describe water as being fluid and steel as being nonfluid because of the way they respond when we handle them. "Handling" introduces deforming forces.

When a deforming force is applied to a material body it is found that there are generally four major types of mechanical responses, as illustrated schematically in Fig.1.2.1. In the first type the body does not deform at all, no matter how large the applied force. A body of this type is referred to as being "rigid." In the second type the body deforms under the action of the force but it regains its initial geometry when the force is removed. It is referred to as an "elastic" body. In the third type the body deforms under the action of the force but it remains in its deformed state when the force is removed. It is referred to as a "plastic" body. In the fourth type, finally, the body deforms under the action of the force and *continues to deform when the force is removed.* It is referred to as a fluid body.

Examples of elastic solids are steel, rubber, and wood, all of which deform under the action of a force and then regain their shape when the force is removed. The difference between rubber and steel is one of degree, not of type. Steel requires much larger force than does rubber to be deformed, but it does deform. Steel is much less elastic than rubber, but it is elastic nevertheless; hence, both are in the same category of elastic solids. So are many other materials, with a wide range of elasticity. In fact even rigid solids can be included in this category as a limiting case of the elasticity range in which the force required to produce deformation is infinitely large. This view is useful because fluids can then also be included as a limiting case at the other end of the elasticity range where the force required to produce deformation is infinitely small. A fluid body not only continues to deform after the deforming force has been removed *but the force required*

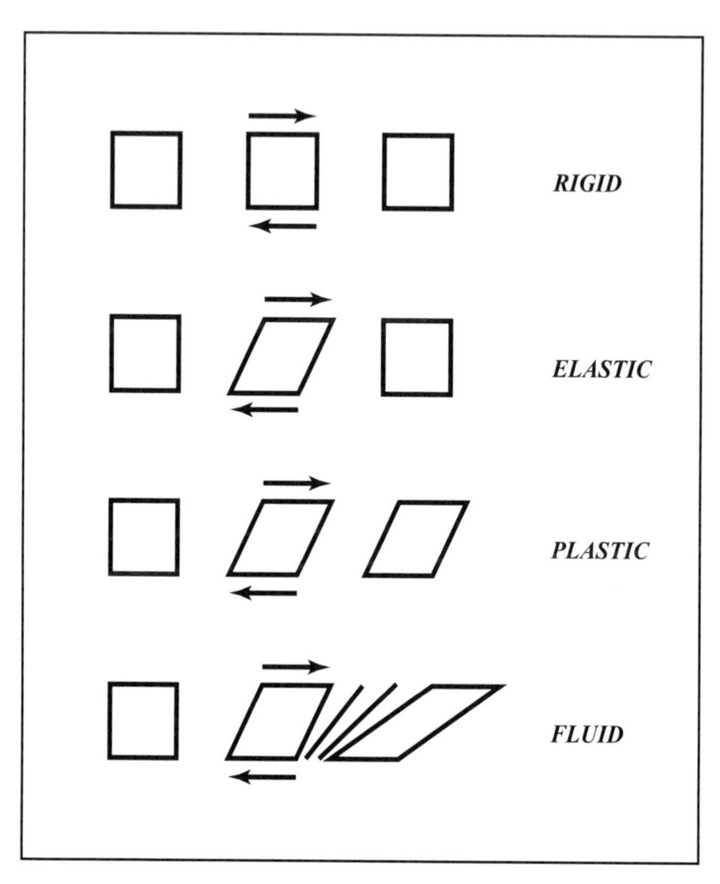

FIGURE 1.2.1. Mechanical classification of material bodies. An initially rectangular body (left column) is subjected to a shearing force (middle column), then the force is removed (right column). A rigid body does not deform at all under the action of the force. An elastic body deforms under the action of the force but regains its shape when the force is removed. A plastic body deforms under the action of the force and remains in the deformed state when the force is removed. A fluid body deforms under the action of the force and continues to deform when the force is removed.

to produce the initial deformation is close to zero– Anything above zero, however small. This is a second identifying property of fluids.

As a result of this unique property, a fluid body cannot support a nonzero deforming force while remaining at rest, however small the force is, because as soon as the force is applied the body will deform and then continue to deform. An elastic solid, by contrast, can remain at rest in its deformed state, in equilibrium against the deforming force.

1.3 Microscopic and Macroscopic Scales

The inability of fluids to be in equilibrium with a deforming force, their inability to oppose that force while remaining at rest, is a remarkable property that has its origin in the molecular structure of fluids. We recall that all matter is molecular in structure, on a scale that we shall refer to as the "microscopic" scale. The mechanical behavior discussed in the previous section and the dynamics of fluids in general, on the other hand, usually occur on a much larger scale that we shall refer to as the "macroscopic" scale.

The mechanical behavior of a material body on the *macroscopic* scale, indeed the way in which the body holds itself physically together as a body, depends on conditions that prevail entirely on the *microscopic* scale. More specifically, they depend on a delicate balance between forces of attraction and forces of repulsion between the molecules that make up the body [12,13]. According to current theory, in the solid state forces of attraction predominate to the effect that neighboring molecules cannot break away from each other, thus giving the body its "solidity" and its resistance to deformation. In the fluid state, by contrast, the balance between forces of attraction and repulsion is so precarious that neighboring molecules are almost free to move away from each other, thus giving the body its "fluidity."

Superimposed on this delicate balance of forces is the chaotic vibration that molecules are generally engaged in, in both the solid and fluid states. This makes the microscopic scale totally unsuitable for a study of fluid behavior on the macroscopic scale. Furthermore, on the microscopic scale material bodies are highly discontinuous because of the (relatively) vast empty space between molecules. On the macroscopic scale, by contrast, the same bodies appear continuous, and for all practical and analytical purposes we would like to treat them as such.

These difficulties are resolved in the study of fluid dynamics by working only on the macroscopic scale and by treating a fluid body as consisting not of molecules but of small pieces which are *continuous with each other, with no empty space in between* (Fig.1.3.1). In fluid dynamics these pieces are usually referred to as "fluid elements."

The validity of this continuum view of fluids as a working model rests on the requirement that a fluid element be both very large on the microscopic scale and very small on the macroscopic scale. The first is required so that a fluid element contains a sufficiently large number of molecules to act as a fluid body, the second is required so that the element can be treated as the smallest possible object, in fact a "point", on the macroscopic scale. While they seem contradictory, these requirements are actually easy to satisfy in practice for most fluids under normal conditions of temperature and pressure. As an example, one cubic millimeter of air under standard temperature and pressure contains approximately 10^{16} molecules. Thus, even

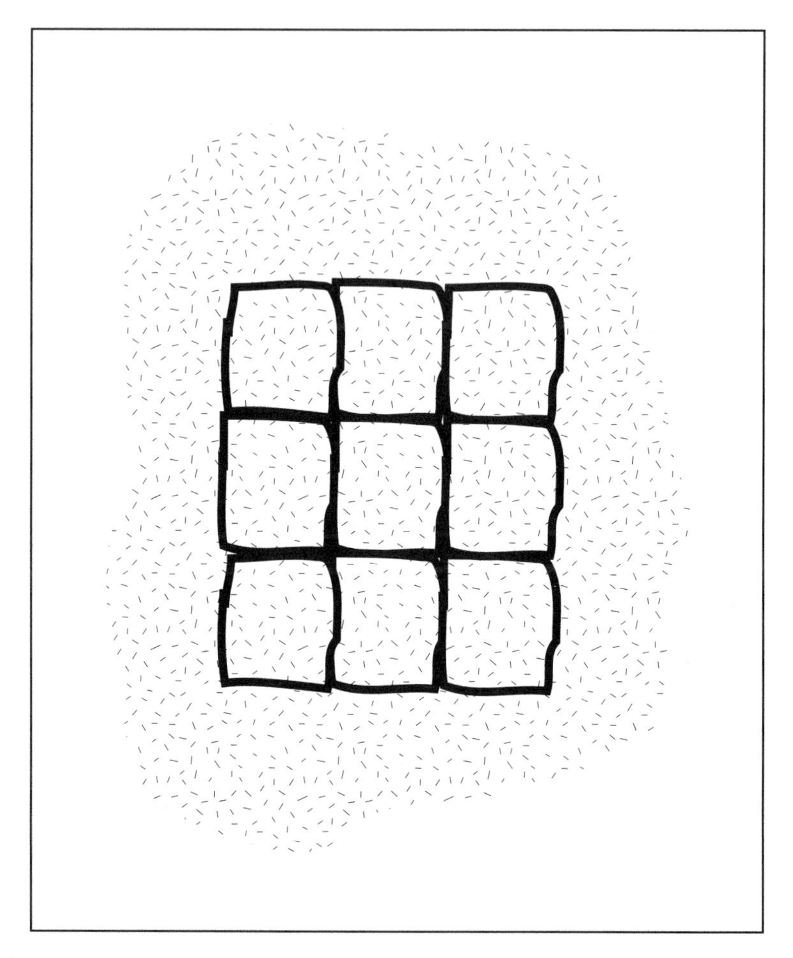

FIGURE 1.3.1. Schematic view of the microscopic and macroscopic scales. On the microscopic scale a fluid body consists of a discontinuous collection of molecules. On the macroscopic scale we view it as a *continuous* collection of elements. The figure depicts the relation between the two scales only in concept, not in proportion, because the difference between the two scales is much too large to be shown in the same figure.

if we consider a fluid element to be one millionth of a cubic millimeter, it would still contain 10^{10} molecules, which is sufficiently large on the microscopic scale for the element to behave as a fluid body, but sufficiently small on the macroscopic scale to be treated as a point.

The continuum view of fluids and the concept of a fluid element are essential in the study of fluid dynamics because they allow the use of continuous functions to describe the properties of a flow field on the macroscopic scale. In that description every "point" represents a fluid element, and properties at that point, such as velocity or density, then represent the properties of a fluid element at that point. We shall see later that neither the identity nor the mass or shape of the element are required in that description.

1.4 What Is Flow?

When a fluid body is in motion, its elements can move in different directions and at different speeds. Because of this, a fluid body rarely moves in bulk as a solid body. More often an element of fluid will be moving faster or slower than its neighboring elements, and because these elements are not truly separate from each other and must remain continuous with each other (Fig.1.3.1), the different velocities produce a shearing motion that is the hallmark of fluid flow [14,15].

A fluid element in a flow field is typically positioned between a faster moving element on one side and a slower moving element on the other, somewhat like a vehicle traveling on a multilane highway. The result of this "velocity gradient" is that the element is in a state of continuous deformation, and so are other elements of the body. The body is in a state of flow. Flow is a state of continuous deformation.

When a cup of coffee is lifted carefully so as not to disturb the stillness of the fluid within, there will be no flow even though all elements of the fluid body are in motion (moving with the cup). There is no flow in the absence of continuous deformation. If a spoon is used to disturb the fluid within the cup, however, fluid elements will move in different directions and at different speeds, a state of continuous deformation will be initiated, a state of flow. Generally there is very little that can be done to a fluid body without causing it to flow, because the force required to initiate the state of continuous deformation is infinitely small.

In a flow field the velocity of fluid elements varies from point to point, that is, the velocity at a point is a function of position within the fluid body, thus producing velocity gradients within the body. A flow field is generally a field of velocity gradients. The gradients may be in all directions, but for the purpose of discussion we may consider only one of these as illustrated in Fig.1.4.1. Here the x-component of velocity u varies in the y direction, producing a gradient du/dy. A fluid element under this gradient will be in a state of continuous shear, a state of continuous deformation as illustrated schematically in Fig.1.4.1.

The most important difference between fluid and elastic bodies is that in the case of an elastic body a force is required to hold the body in a state of deformation, similar to that required to hold a spring in a non-neutral state. In the case of a fluid body a force is required to hold the body in a state of *continuous deformation*, that is, in a state of flow. The characteristic property of fluids that the force must overcome is called their "viscosity." The viscosity of fluids is their intrinsic resistance to *rate of deformation*, in

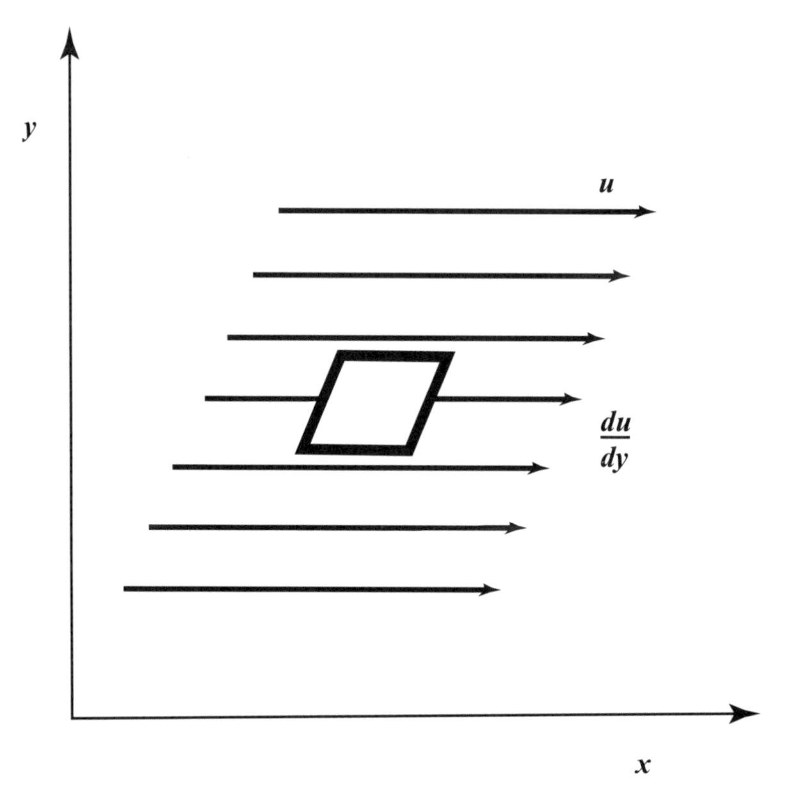

FIGURE 1.4.1. Flow is a state of velocity gradients. Only one gradient is illustrated here, resulting from variation of the x-component of velocity u in the y direction. An element of fluid within this gradient is in a state of continuous deformation, a state of flow.

the same way that the elasticity of elastic solids is their intrinsic resistance to deformation.

The viscous property of fluids at once explains why a force is needed to maintain a state of flow, and how and why a flow, once started, ever stops. A force is required to maintain a state of flow because the viscosity of the fluid opposes the velocity gradients prevailing in that state. And a state of flow that has already been started will gradually diminish if the force sustaining the flow is removed, because the viscosity of the fluid will be unopposed in gradually diminishing any velocity gradient within the flow field.

The resisting force produced by viscosity at a point in a flow field increases with the *rates* of deformation being produced at that point, which, in turn, are represented by the velocity gradients at that point. A compelling demonstration of this is offered when trying to turn the cap of a jar of honey or some other sticky liquid. If a thin layer of honey is trapped between the cap and the jar, as is usually the case, the turning action creates a velocity gradient under which fluid in that layer will be in a state of continuous deformation. If the turning action is very slow, it will meet little resistance. But if the turning action is done with any speed at all, a high resistance will be experienced in proportion to that speed.

1.5 Eulerian and Lagrangian Velocities

When a solid body is in motion, it is usually possible to consider the velocity of the body as a whole. In the case of a fluid body, this is rarely possible because of the ease with which the body can be deformed, and hence the ease, although not complete freedom, with which different elements of the body can move independently of each other. In the case of a fluid body, therefore, motion usually involves a *velocity field* within the body rather than a single velocity. The field represents the aggregate of velocities of all elements of the body, and because these elements are mere "points" on the macroscopic scale, the velocity field represents the aggregate of velocities of all points within the fluid body.

There are two ways in which the velocity field within a fluid body can be mapped [16]. In the first, known as the "Lagrangian" method, the initial position of each element of the body, that is, the position of the element before the motion started, is recorded. Then the motion of each element is followed, individually, in the usual way, like the motion of a single isolated particle. For each element, this velocity will in general be a function of time only, as in the case of an isolated particle, but for all elements of the fluid body the aggregate of all such velocities will be a function of time and of the identity of each element. These so called Lagrangian velocities will be functions of time and of the initial positions of all elements of the fluid body.

In a cylindrical polar system of coordinates x, r, θ (Fig.1.5.1), if we denote the Lagrangian velocity components by U, V, W, respectively, then

$$\begin{aligned} U &= U(x_0, r_0, \theta_0, t) \\ V &= V(x_0, r_0, \theta_0, t) \\ W &= W(x_0, r_0, \theta_0, t) \end{aligned} \tag{1.5.1}$$

where x_0, r_0, θ_0 are initial coordinates of different elements of the body, also known as "material coordinates." Each set of values (x_0, r_0, θ_0) identifies a particular fluid element, and the corresponding velocities would be the velocities of that element. Thus, $U(1, 2, 3, t)$ is the velocity of the element whose position when the motion started was $x_0 = 1$, $r_0 = 2$, $\theta_0 = 3$. While this scheme seems logical, it is in fact highly impractical because it requires that the identity of each element of a fluid body is noted and followed at all time, which is neither easy to achieve nor actually useful for the purpose of mapping a flow field.

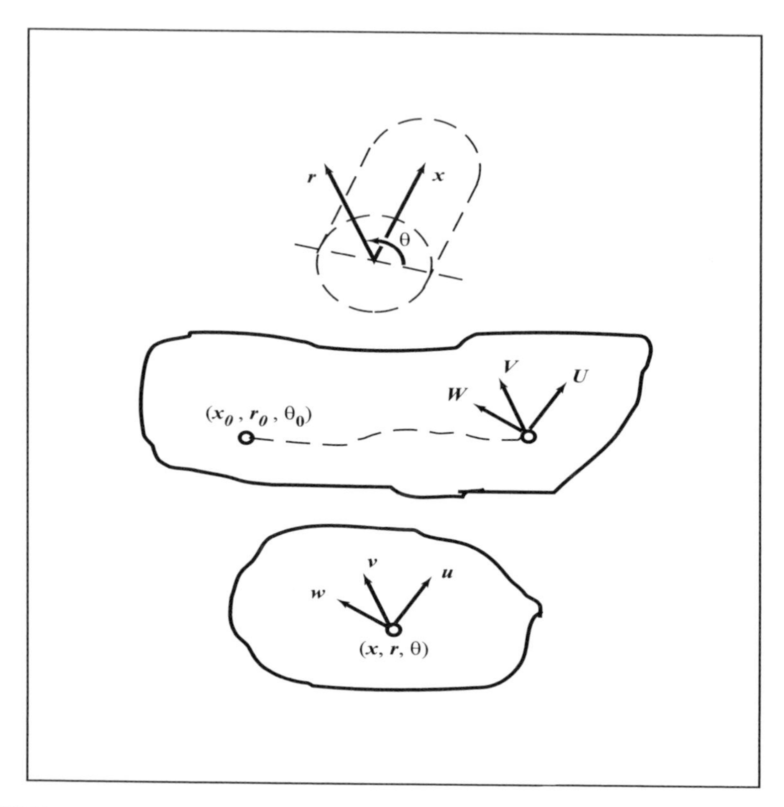

FIGURE 1.5.1. Cylindrical coordinates (x, r, θ). Lagrangian and Eulerian velocities $(U, V, W;\ u, v, w)$. The Lagrangian velocities in a flow field represent the velocities of *identifiable* fluid elements, in terms of their initial positions (x_0, r_0, θ_0). The Eulerian velocities represent the velocities recorded at specific coordinate *positions* (x, r, θ) within the flow field, therefore representing the velocities of different elements that occupy these positions at different times.

The second way in which the velocity field within a fluid body can be mapped, known as the "Eulerian method," is based not on velocities of identifiable elements of the body but on velocities recorded at identifiable coordinate positions within the body. These so called Eulerian velocities are thus functions of time and of *coordinate positions* within the body, not of initial positions of fluid elements. If we denote these by u, v, w, then

$$\begin{aligned} u &= u(x, r, \theta, t) \\ v &= v(x, r, \theta, t) \\ w &= w(x, r, \theta, t) \end{aligned} \tag{1.5.2}$$

Here $u(1, 2, 3, t)$ is the velocity measured at position $x = 1$, $r = 2$, $\theta = 3$ in the flow field, at time t. Since at different times this coordinate position within the body is occupied by different fluid elements, Eulerian velocities at a point in a flow field do not represent the velocities of the *same* fluid element at all time. They represent the velocities of different elements that occupy this position over time.

While Eulerian velocities seem less logical and less easy to interpret, they are the velocities of choice in most fluid flow problems because their practicality far outweighs the conceptual and analytical difficulties that they entail. The practicality results from the ease with which the velocities can be measured at specific coordinate positions within a flow field, by simply placing instruments at the required positions. And the analysis of fluid flow problems are more meaningful in terms of Eulerian velocities because in practice it is these velocities that we would normally be interested in: the velocities at this or that point in a flow field, rather than the velocities of this or that identifiable fluid element.

1.6 Acceleration in a Flow Field

The acceleration of an object in motion is the rate of change of its velocity with time [17,18]. When the object is a fluid element within a flow field, its velocity is given appropriately by the Lagrangian velocity components, because these are functions of the identity of the fluid element, that is, functions of its material coordinates. Thus, as before, if in a cylindrical polar system of coordinates x, r, θ the Lagrangian velocity components are denoted by $U(x_0, r_0, \theta_0, t)$, $V(x_0, r_0, \theta_0, t)$, $W(x_0, r_0, \theta_0, t)$, where x_0, r_0, θ_0 are material coordinates and t is time, then the acceleration components in the directions of x, r, θ are given by

$$a_x = \frac{\partial U}{\partial t}, \quad a_r = \frac{\partial V}{\partial t}, \quad a_\theta = \frac{\partial W}{\partial t} \tag{1.6.1}$$

Because U, V, W are functions of x_0, r_0, θ_0, t, their partial derivatives with respect to t are obtained by keeping x_0, r_0, θ_0 constant, which appropriately means that the acceleration being evaluated is that of the particular element identified by the material coordinates (x_0, r_0, θ_0).

As before, however, while the above seems logical, it is impractical because it requires use of the Lagrangian velocity components, which, in turn, requires a continuous record of the identity of fluid elements. For practical purposes we would like to express the acceleration components in terms of the Eulerian velocities u, v, w, instead of the Lagrangian velocities. But because u, v, w are functions of *position coordinates* x, r, θ and time t, their partial derivatives with respect to time, $\partial u/\partial t, \partial v/\partial t, \partial w/\partial t$, are obtained by keeping x, r, θ constant. These derivatives therefore represent the rate of change of velocities at certain *locations* within the flow field, not of certain fluid elements. And these locations are occupied by different fluid elements at different times.

These difficulties can be resolved if we now take (x, r, θ) to represent not a fixed coordinate position within the flow field but the *instantaneous position at time t of a fluid element in motion*, which makes x, r, θ functions of t. Thus if we write Eq.1.5.2 as

$$
\begin{aligned}
u &= u\{x(t), r(t), \theta(t), t\} \\
v &= v\{x(t), r(t), \theta(t), t\} \\
w &= w\{x(t), r(t), \theta(t), t\}
\end{aligned}
\tag{1.6.2}
$$

then we now see the Eulerian velocities not as the velocities at location x, r, θ, but as the velocities of the fluid element that happens to occupy that location at time t. And the acceleration of that fluid element is now given appropriatley by the *total* derivatives of these velocities with respect to time, that is,

$$
\begin{aligned}
a_x &= \frac{Du}{Dt} \\
&= \frac{\partial u}{\partial t} + \frac{\partial u}{\partial x}\frac{dx}{dt} + \frac{\partial u}{\partial r}\frac{dr}{dt} + \frac{\partial u}{\partial \theta}\frac{d\theta}{dt} \\
a_r &= \frac{Dv}{Dt} - \frac{w^2}{r} \\
&= \frac{\partial v}{\partial t} + \frac{\partial v}{\partial x}\frac{dx}{dt} + \frac{\partial v}{\partial r}\frac{dr}{dt} + \frac{\partial v}{\partial \theta}\frac{d\theta}{dt} - \frac{w^2}{r} \\
a_\theta &= \frac{Dw}{Dt} + \frac{vw}{r} \\
&= \frac{\partial w}{\partial t} + \frac{\partial w}{\partial x}\frac{dx}{dt} + \frac{\partial w}{\partial r}\frac{dr}{dt} + \frac{\partial w}{\partial \theta}\frac{d\theta}{dt} + \frac{vw}{r}
\end{aligned}
\tag{1.6.3}
$$

The added terms in the expressions for a_r and a_θ arise because of curvature in the cylindrical system [19,20]. Furthermore, since $x(t), r(t), \theta(t)$ are the instantaneous coordinates of the fluid element at x, r, θ at time t, then their derivatives with respect to t represent the Eulerian velocity components at that location and that instant in time, that is,

$$
\frac{dx}{dt} = u, \quad \frac{dr}{dt} = v, \quad r\frac{d\theta}{dt} = w \tag{1.6.4}
$$

and the expressions for the acceleration components finally become

$$
\begin{aligned}
a_x &= \frac{\partial u}{\partial t} + u\frac{\partial u}{\partial x} + v\frac{\partial u}{\partial r} + \frac{w}{r}\frac{\partial u}{\partial \theta} \\
a_r &= \frac{\partial v}{\partial t} + u\frac{\partial v}{\partial x} + v\frac{\partial v}{\partial r} + \frac{w}{r}\frac{\partial v}{\partial \theta} - \frac{w^2}{r} \\
a_\theta &= \frac{\partial w}{\partial t} + u\frac{\partial w}{\partial x} + v\frac{\partial w}{\partial r} + \frac{w}{r}\frac{\partial w}{\partial \theta} + \frac{vw}{r}
\end{aligned}
\tag{1.6.5}
$$

It is important to note that the partial derivatives $\partial u/\partial t, \partial v/\partial t, \partial w/\partial t$, are rates of change of u, v, w with respect to time, *keeping* x, r, θ *constant*, and that these derivatives do not represent the components of acceleration of the fluid element at x, r, θ. They represent only part of that acceleration. The total acceleration of the element in terms of Eulerian velocities depends on the partial derivatives of these velocities with respect to time as well as on the *velocity gradients* at that point. In short, the acceleration is given by the *total* derivatives of these velocities with respect to time, also known as their *convective* derivatives.

1.7 Is Blood a Newtonian Fluid?

The way in which fluids resist rates of deformation is a fundamental property of the equations that govern their behavior. If for simplicity we consider rate of deformation in only one direction, as illustrated in Fig.1.4.1, then at issue is the relation between the shear stress τ, which represents the resisting force, and the corresponding velocity gradient du/dy, which represents the rate of deformation in this case. The relation most commonly used is a linear one, with a constant of proportionality μ known as the coefficient of viscosity and whose value is a characteristic property of the fluid, that is,

$$\tau = \mu \frac{du}{dy} \tag{1.7.1}$$

This relation was first derived by Newton, and fluids whose behavior is consistent with it are referred to as Newtonian fluids [21,22].

Many common fluids are found to behave as Newtonian fluids, among them air, water, and oil. Others may behave as Newtonian fluids when the rates of deformation and hence velocity gradients within a flow field are small, and as non-Newtonian fluids when the gradients are large. Theoretical studies have concluded, in fact, that the linear relation in Eq.1.7.1 is only an approximation for small rates of deformation, but one that has a wide range of validity. Other relations have been explored in both theoretical and experimental studies.

The question of whether blood is a Newtonian fluid is a long standing one [23]. Indeed, the corpuscular nature of whole blood raises the question of whether it can be treated as a continuum, and the peculiar makeup of plasma makes it *seem* different from more common fluids. There is no doubt that blood cannot be treated as a Newtonian fluid in general and under all circumstances. What is a more meaningful question is whether blood can be treated as a Newtonian fluid in the study of a particular blood-flow problem. It is meaningful to consider only individually, in each case, whether or not non-Newtonian effects play a significant primary role in the phenomenon being studied.

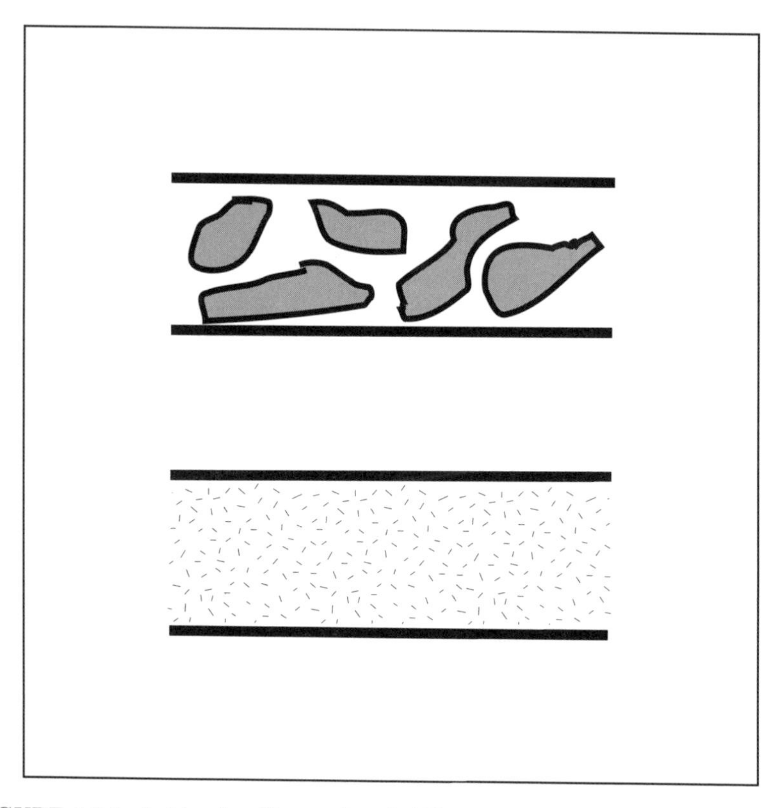

FIGURE 1.7.1. Is blood a Newtonian fluid? The question is tied much to the diameter of the tube in which blood is flowing. When the diameter is comparable with the scale of corpuscular structure of blood (top), the assumption of a Newtonian fluid, indeed the assumption of blood as a continuum, is clearly untenable. However, when the tube diameter is large on the corpuscular scale (bottom), both assumptions are found to be fairly adequate.

A great deal of understanding, indeed the overwhelming majority of what we know to date about the dynamics of blood flow, has been achieved by treating blood as a Newtonian fluid, or more accurately by using Newtonian relations between shear stresses and velocity gradients. This is not to say that blood *is* a Newtonian fluid, but only that the Newtonian relations have proved adequate for much of what has been studied so far. There is no doubt that under extreme conditions that produce unusually high velocity gradients, as in space or deep sea travel, or accidental impact, it may become necessary to take the secondary effects of non-Newtonian behavior into account. Also, near the capillary level of the vascular tree where vessel diameters become comparable with the size of the discrete corpuscles of blood (Fig.1.7.1), the continuum model of blood and its assumed Newtonian behavior become clearly inadequate. But in the core of more common blood flow problems, including the global dynamics of the circulation and of the heart, and the local dynamics of flow around obstacles and through vascular junctions, the Newtonian model has proved adequate so far.

In particular, a fundamental understanding of the dynamics of pulsatile flow has been achieved on the basis of Newtonian relations between shear stresses and velocity gradients, although some studies have considered the problem of pulsatile flow of non-Newtonian fluids. In this book we shall be concerned with the basic characteristics of pulsatile flow of Newtonian fluids only. These characteristics provide the basic understanding required in order to deal with more complicated cases. Indeed they provide a necessary reference that results in more complicated cases can be compared with.

1.8 No-Slip Boundary Condition

A fundamental consequence of the viscous property of fluids is that in fluid flow there can be no "step" change in velocity at any point within the flow field. The reason for this is that the velocity gradient at a point is a measure of the rate of deformation of the fluid element at that point, which is resisted by the viscosity of the fluid. A force is therefore required to maintain the velocity gradient, and the higher the gradient, the higher the force, as evident from Eq.1.7.1. A step change in velocity at a point in a flow field implies that the velocity gradient is infinite at that point, which is not possible because the force required to maintain it would also have to be infinite.

In particular, at the interface between a fluid and a solid boundary (Fig.1.8.1), as at the inner wall of a tube, the velocity of the fluid in contact with the wall must be the same as the velocity of the wall, otherwise there would be a step change in velocity at that point. This so-called "no-slip boundary condition" is a fundamental condition that must be satisfied in any analysis of viscous flow. For flow in a tube, it implies that fluid does not simply "slide" down the tube as a bullet or a bolus. Instead, fluid in contact with the tube wall does not move at all, because it must have zero velocity relative to the wall, and fluid further and further away from the wall moves with velocities commensurate with distance from the wall, fluid along the axis of the tube having the highest velocity. The highest velocity at the center of the tube and zero velocity at the tube wall must be connected by a smooth profile with no step change at any point.

The condition of no-slip at the tube wall is the reason for which flow in a tube requires pumping power to maintain. In the absence of this condition, fluid would be able to slide down the tube as a bolus, and under steady flow conditions no energy would be required to drive it. In the presence of no-slip, pumping power is required in order to maintain the inevitable velocity gradient at the tube wall. The amount of power increases as the flow rate increases, because this increases the velocity gradient at the wall.

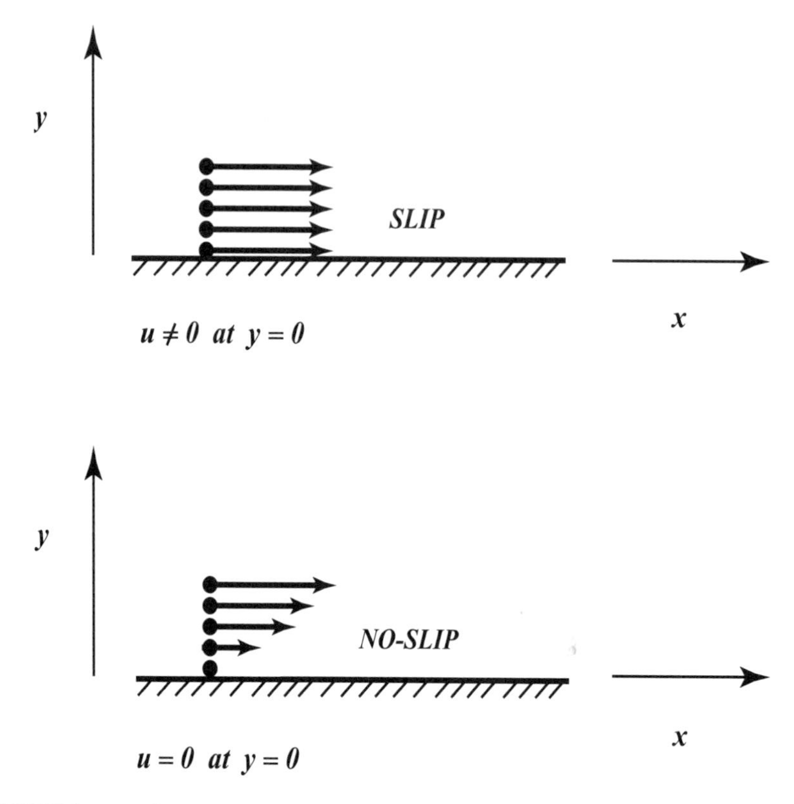

FIGURE 1.8.1. Condition of no–slip at a solid fluid interface. Viscosity does not allow a fluid to slip past a solid boundary. Fluid in contact with the boundary must have zero velocity, and the velocities of neighboring elements must change smoothly to meet that value at the wall.

The power is also higher for higher coefficients of viscosity, more viscous fluids require higher pumping power to drive through a tube.

The possibility of partial slip between blood and endothelial tissue at the inner wall of a blood vessel has been explored in some studies in the past in an attempt to explain an apparent drop in the viscosity of blood as it moves into vessels of small diameter. The phenomenon is known as the Fahraeus–Lindqvist effect [24–27]. Partial slip would reduce the amount of pumping power required to maintain a given flow rate q through the tube and is therefore equivalent to a reduction in the coefficient of viscosity as explained above. Slip or partial slip have not been directly demonstrated, however. Early observations noted the "skidding" of red blood cells along the vessel wall, and this was thought at first to indicate slip, but it soon became clear that skidding must have occurred over a thin layer of plasma, acting as a lubricant. The condition of no-slip therefore pertained to the plasma in contact with the vessel wall rather than to the observed cells.

Direct measurements of velocity profiles within blood vessels have so far tended to support a condition of no-slip at the tube wall.

1.9 Laminar and Turbulent Flow

In late 19th century Osborne Reynolds made one of the most important discoveries about the flow in a tube [21,22]. In a series of experiments designed to study the basic characteristics of the flow, Reynolds rendered the flow visible by injecting dye at the tube's entrance, then changed the flow rate to see how this affected what he observed. The injection of dye had the effect of "marking" elements of fluid so that their subsequent course could be observed.

Reynolds found that at low flow rates the marked elements produced streaklines which were fairly distinct and ran parallel to the axis of the tube. At higher flow rates, however, the streaklines became increasingly unstable, eventually braking down and causing the dye to diffuse over the whole cross section of the tube, streaklines being no longer distinct or visible. Reynolds identified what he observed as two different types of flow. Today they are known widely as "laminar" and "turbulent" flow.

Later, more advanced technology, which was not available to Reynolds, showed that in laminar flow fluid elements move only in the direction of the flow, in the direction of the Eulerian velocity vector at each point (Fig.1.9.1). In turbulent flow, by contrast, fluid elements vibrate randomly in all directions, at high frequency and with small amplitude, as they move in the main flow direction. Thus, in turbulent flow a fluid element typically has a "mean" velocity component in the main flow direction, plus small-amplitude oscillatory velocity components in all directions. The situation is a remarkable revisiting of the random motion of molecules on the microscopic scale, which is dealt with by treating a fluid body as a continuum of fluid elements on the much larger macroscopic scale. The difficulty with turbulence is that the scale on which it occurs is already the larger macroscopic scale, and it is the fluid elements themselves that are now engaged in random motion.

Reynolds found that the onset of turbulence depends not only on the average flow velocity through the tube, $\overline{u}$, but also on the density ρ and viscosity μ of the fluid, and diameter d of the tube. His most important contribution was to then recognize that the onset of turbulence actually depends not on $\overline{u}, \rho, \mu, d$ individually but on the nondimensional combination

$$R = \frac{\rho \overline{u} d}{\mu} \tag{1.9.1}$$

which today is known universally and appropriately as the Reynolds number. Reynolds' experiments suggested that transition from laminar to

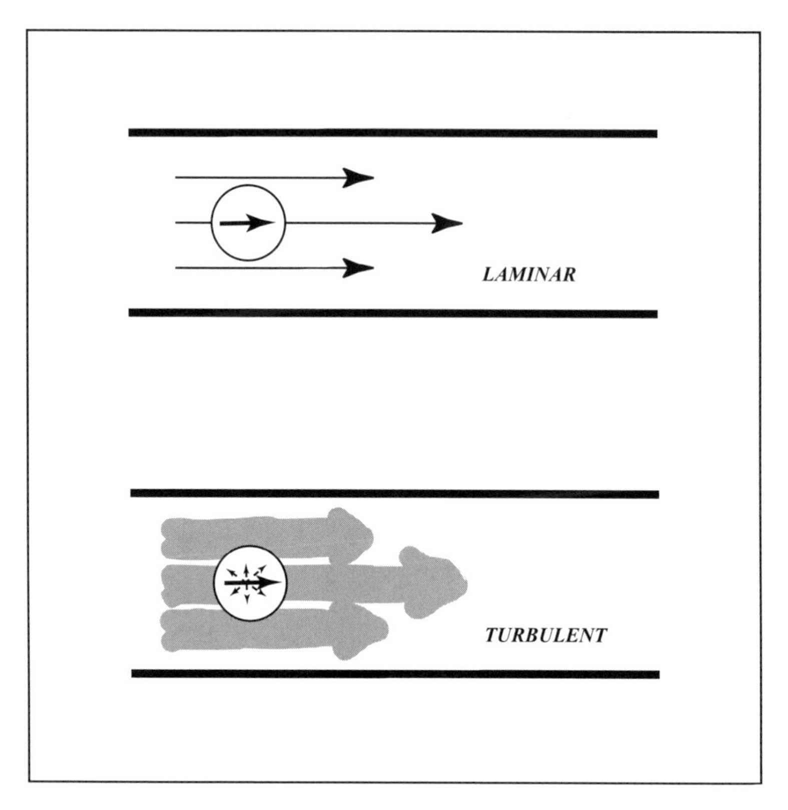

FIGURE 1.9.1. Laminar and turbulent flow. In laminar flow, fluid elements move only in the main flow direction. In turbulent flow, fluid elements vibrate randomly in all directions as they move in the main flow direction.

turbulent flow occurred at $R \approx 2,000$, but, since then, it has been found that the transition can be delayed to much higher values of R if flow disturbances at the entrance to the tube and surface roughness at the tube wall are kept to a minimum. Present understanding is that the value $R = 2,000$ is a "lower bound" below which flow will remain in the laminar state even if disturbed. At higher values of R the flow becomes increasingly unstable and *may* become turbulent depending on prevailing destabilizing conditions.

In blood flow, the highest flow velocities occur in the aorta as it leaves the heart for distribution to the rest of the body. Assuming steady flow at first, in a human aorta of approximately 2.5 cm in diameter and an average cardiac output of 5 L/min, the average velocity is given by

$$\overline{u} = \frac{5,000}{\pi(\frac{2.5}{2})^2 \times 60} \approx 17 \; cm/s \qquad (1.9.2)$$

Taking the density $\rho \approx 1\, g/cm^3$ and viscosity $\mu \approx 0.04\, g/cm\ s$, we obtain

$$R = \frac{\rho \bar{u} d}{\mu} = \frac{1 \times 17 \times 2.5}{0.04} \approx 1,063 \tag{1.9.3}$$

indicating that under these assumptions the Reynolds number is well below 2,000.

In *pulsatile* flow, at the peak of the oscillatory cycle, and under conditions of higher cardiac output, this value of R can be exceeded considerably, thus the possibility exists for turbulent flow to occur at this level of the arterial tree and in part of the oscillatory cycle. Turbulent flow may also occur locally as blood flows through diseased blood vessels and heart valves, sometimes producing so called "murmurs." At higher levels of the vascular tree, however, the Reynolds number diminishes rapidly because average flow velocities, vessel diameters, and ocillatory flow peaks, all diminish rapidly, leaving conditions under which laminar flow is stable. For this reason much of the work on pulsatile flow deals with laminar flow, and we follow this practice in this book.

1.10 Problems

1. A body of sand appears to flow when poured: Is sand a fluid? Give reasons for your answer. Compare and contrast the granular structure of sand with the corpuscular structure of blood.
2. Discuss blood flow in a tube when the tube diameter is less than 10 microns as in a capillary, and when the tube diameter is about 25 mm as in the aorta.
3. There is very little that can be done to a fluid body without causing it to flow. Can you identify a situation in which a fluid body is moving, yet not flowing?
4. The velocity profile in Poiseuille flow is a graphical presentation of the velocity not of specific elements of the fluid but at specific *positions* in a cross section of the tube (Fig.3.4.1). Discuss the meaning of this so-called "Eulerian" description of the flow field within a tube, and the reason for using it.
5. True or false?

   a. The acceleration in a flow field is zero when the Eulerian velocities are not functions of time.
   b. The acceleration in a flow field is zero when the Eulerian velocities are not functions of position.
   c. The acceleration in a flow field is zero when the flow is steady.
   d. The acceleration in a flow field is zero when the flow is uniform.

6. When the flow in a blood vessel was first visualized under the microscope, red blood cells were observed to "skid" along the vessel wall in

a manner which suggested that the condition of no–slip may not necessarily be satisfied at the vessel wall as it is at the wall of a tube. Discuss the ramifications of this suggestion and the likelihood of its validity.

7. Define the Reynolds number R for flow in a tube and explain the significance of the value $R = 2,000$. Where is this value of the Reynolds number likely to be met in the cardiovascular system?

1.11 References and Further Reading

1. LaBarbera M, Vogel S, 1982. The design of fluid transport systems in organisms. American Scientist 70:54–60.
2. LaBarbera M, 1990. Principles of design of fluid transport systems in zoology. Science 249:992–1000.
3. LaBarbera M, 1991. Inner currents: How fluid dynamics channels natural selection. The Sciences Sept/Oct:30–37.
4. West GB, Brown JH, Enquist BJ, 1997. A general model for the origin of allometric scaling laws in biology. Science 276:122–126.
5. De Robertis EDP, Nowinski NW, Saez FA, 1970. Cell Biology. Saunders, Philadelphia.
6. Lyall F, El Haj AJ, 1994. Biomechanics and Cells. Cambridge University Press, Cambridge.
7. Congdon ED, 1922. Transformation of the aortic arch system during the development of the human embryo. Contributions to Embryology, Carnegie Institute of Washington 14:47–110.
8. Zamir M, Sinclair P, 1990. Continuum analysis of the common branching patterns in the human arch of the aorta. Anatomy and Embryology 181:31–36.
9. Frankel JP, 1957. Principles of the Properties of Materials. McGraw–Hill, New York.
10. Long RR, 1961. Mechanics of Solids and Fluids. Prentice-Hall, Englewood Cliffs, New Jersey.
11. Eisenstadt MM, 1971. Introduction to Mechanical Properties of Materials. Macmillan, New York.
12. Batchelor GK, 1967. An Introduction to Fluid Dynamics. Cambridge University Press, Cambridge.
13. Tabor D, 1991. Gases, Liquids, and Solids: and Other States of Matter. Cambridge University Press, Cambridge.
14. Van Dyke M, 1982. An Album of Fluid Motion. Parabolic Press, Stanford, California.
15. Nakayama Y, 1990. Visualized Flow. Pergamon Press, Oxford.
16. Truesdell C, Toupin RA, 1960. The classical field theories. In: Flugge S (ed), Encyclopedia of Physics, Vol. III/1: Principles of Classical Mechanics and Field Theory. Springer-Verlag, Berlin.
17. Meriam JL, 1966. Dynamics. John Wiley, New York.

18. Chorlton F, 1969. Textbook of Dynamics. Van Nostrand, Princeton, New Jersey.
19. Moon PH, Spencer DE, 1961. Field Theory for Engineers. Van Nostrand, Princeton, New Jersey.
20. Curle N, Davies HJ, 1968. Modern Fluid Dynamics: I. Incompressible Flow. Van Nostrand, Princeton, New Jersey.
21. Rouse H, Ince S, 1957. History of Hydraulics. Dover Publications, New York.
22. Tokaty GA, 1971. A History and Philosophy of Fluidmechanics. Foulis, Henley-on-Thames, Oxfordshire.
23. Bergman LE, DeWitt KJ, Fernandez RC, Botwin MR, 1971. Effect of non-Newtonian behaviour on volumetric flow rate for pulsatile flow of blood in a rigid tube. Journal of Biomechanics 4:229–231.
24. Fahraeus R, Lindqvist T, 1931. The viscosity of the blood in narrow capillary tubes. American Journal of Physiology 96:562–568.
25. Cokelet GR, 1966. Comments on the Fahraeus–Lidquist effect. Biorheology 4:123–126.
26. Nubar Y, 1971. Blood flow, slip, and viscometry. Biophysical Journal 11:252–264.
27. Zamir M, 1972. Blood flow, slip, and viscometry. Biophysical Journal 12:703–704.

2

Equations of Fluid Flow

2.1 Introduction

Equations governing steady or pulsatile flow in a tube are a highly simplified form of the equations that govern viscous flow in general. The laws on which the general equations are based and the assumptions by which the simpler equations are obtained define the range of validity of these equations and of any results derived from them. This chapter is therefore devoted to a brief outline of the way in which these equations are derived.

There are many excellent books in which the equations of fluid dynamics are considered more generally and in greater detail. A small selection of these, relevant particularly to the material of this chapter, are listed at the end of the chapter for reference and further reading [1–9].

The material of this chapter is important not only because of the assumptions on which the governing equations are based, but also because the process of deriving these equations enlarges on issues concerning the mechanics of fluids discussed in Chapter 1. In the present chapter we see how these issues are dealt with mathematically.

2.2 Equations at a Point

Equations governing flow in a tube and those governing fluid flow in general are "point equations" in the sense that they apply at individual points within a fluid body rather than to the body as a whole. We recall from

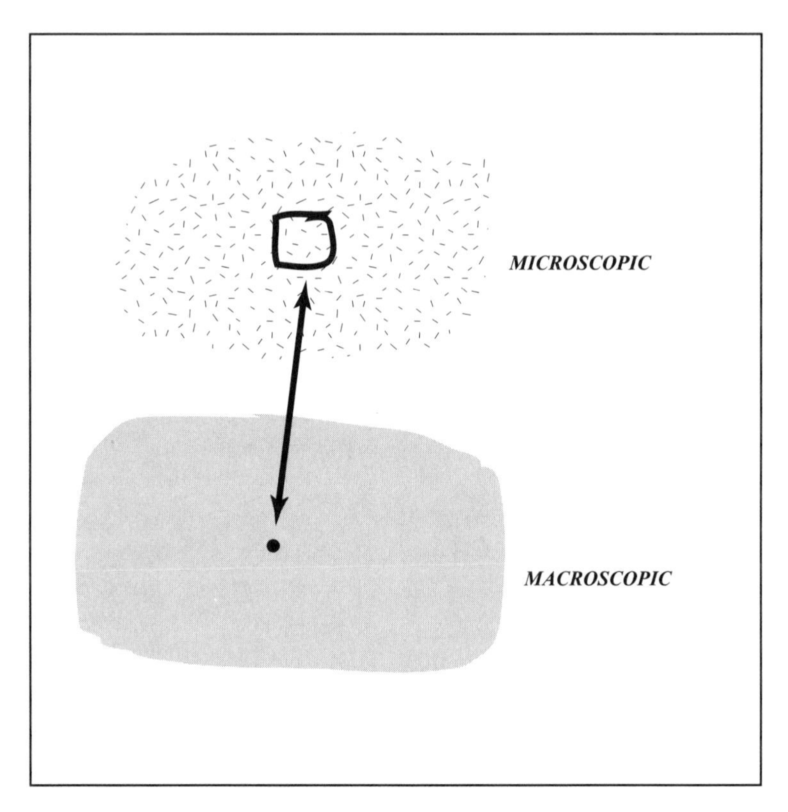

FIGURE 2.2.1. A material element on the microscopic scale becomes a "point" on the macroscopic scale. Equations governing the mechanics of fluid elements therefore become "point equations" on the macroscopic scale. They govern conditions at every point in a flow field, on the macroscopic scale.

discussion in Chapter 1 that on the macroscopic scale these points are *material points*, which actually represent elements of the fluid body, as illustrated schematically in Fig.2.2.1. Equations governing fluid flow are therefore *equations governing the mechanics of fluid elements.*

One of the most important challenges in the formulation of these equations is that fluid elements must be treated in a fairly generic manner. Neither their identity nor their mass, size, or shape is known or specified. Instead, a "presumed" mass m of a fluid element is divided by a "presumed" volume v, to yield density ρ, and, because the fluid element is represented by a point on the macroscopic scale, the density obtained in this way is the *density at a point.*

The concept of density at a point is central to the derivation of equations governing the mechanics of fluids, first because it makes it possible to deal with fluid elements in a generic manner without specifying their mass or volume, and second because it provides an important link between the macroscopic and microscopic scales. Thus the density ρ at a point (x, r, θ)

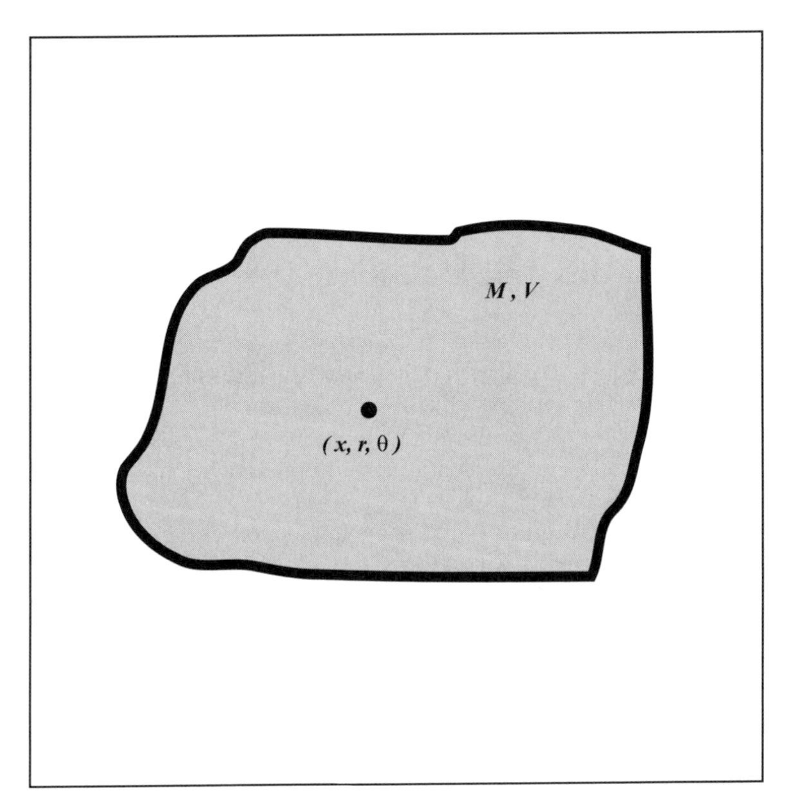

FIGURE 2.2.2. Meaning of "density at a point." A body of fluid of mass M and volume V surrounding the point (x, r, θ) is shrunk to the size of a fluid element at that point, which on the macroscopic scale is represented by that point. In the limit, as the body is shrunk, the density M/V becomes the density of the fluid element at (x, r, θ) and hence the density at that point. The density at a point is thus the density of the fluid element at that point.

in a flow field is the average density of the "bag of molecules" which that material point represents on the macroscopic scale. It is defined not by counting the molecules in the bag, but by considering a much larger bag of molecules and allowing it to shrink to the size of a material point, which is the limit to which the bag can shrink on the macroscopic scale.

Thus considering a volume V and mass M of fluid surrounding the point (x, r, θ) on the macroscopic scale, we define the density at that point by taking the limit of the ratio M/V as this larger body shrinks to the size of a fluid element (Fig.2.2.2), that is,

$$\rho(x, r, \theta) = \lim_{V \to 0} \left(\frac{M}{V} \right) = \frac{m}{v} \tag{2.2.1}$$

The limit is written as $V \to 0$ although it is intended to mean $V \to v$. This is justified on the grounds that v, the volume of a fluid element, is zero on the macroscopic scale. On the macroscopic scale a limit to a point is a limit to a fluid element at that point. Equations governing the mechanics of fluids are point equations in the sense that they govern the mechanics of fluid elements at different points within a flow field.

2.3 Equations and Unknowns

Equations of fluid flow are based on laws that govern the mass and momentum of fluid elements. The first requires that the mass of a fluid element be conserved, the second requires that its momentum obeys Newton's laws of motion. Because mass is a scalar quantity and momentum is a vector quantity, the first of these requirements provides one equation and the second provides three equations.

The total of four governing equations allows the choice of four unknowns. This choice must be made with a view to the type of variables that would be most useful in practice. The choice usually made is that of pressure p and three velocity components, and for reasons discussed in Chapter 1 the latter are chosen to be not the Lagrangian but the Eulerian velocity components u, v, w. The set of four unknowns, p, u, v, w, are thus the dependent variables in the governing equations, and because each of these is a function of position x, r, θ and time t, this basic choice of variables provides a mapping of the velocity and pressure at every point within a flow field and at every point in time.

This choice of variables is favored not only because it provides a good description of the flow field but also because these variables are the most accessible, easiest to measure, in practice. By contrast, the law of conservation of mass and Newton's laws of motion deal with the *mass* of a fluid element, its *acceleration* and the *forces* acting on it. These variables are not easily accessible in practice and they do not provide a good description of the flow field. We shall see later that the laws can be expressed in terms of the favored variables p, u, v, w, instead, although at the price of making the equations more complicated analytically as we saw in Section 1.6. A major challenge in the construction of equations governing fluid flow is how to express laws governing mass, acceleration, and force, in terms of pressure and velocities.

The preceding discussion assumes that the fluid density ρ and viscosity μ are constant and do not vary from point to point in the flow field or with time. When this is not the case, additional equations must be provided to complete the system of equations and unknowns, which may include a thermodynamic equation of state or an energy equation. In this book we shall deal with the basic set of four equations and four unknowns only, as these serve adequately the purpose of description and analysis of pulsatile flow.

2.4 Conservation of Mass: Equation of Continuity

To apply the law of conservation of mass at a point in a flow field we consider a fixed volume of the field in the form of a closed box surrounding the point, then allow the box to shrink to that point. In a cylindrical polar coordinate system the natural choice for the geometry of the box is that produced by incremental change of coordinates $\delta x, \delta r, \delta\theta$, as shown in Fig.2.4.1. The volume of the box is approximately given by

$$V \approx r\delta x\delta r\delta\theta \tag{2.4.1}$$

and the mass contained within this volume is

$$M \approx \rho V \tag{2.4.2}$$

where ρ is density, which may vary within the box, hence the equality is only approximate.

Changes in this mass occur as flow enters and leaves different sides of the box. Because these sides were chosen to coincide with coordinate planes, each side is normal to only one velocity component, the other two being tangential to it and hence do not contribute to mass flow through it.

Thus mass flows entering three sides of the box are as shown in Fig.2.4.1. If mass flows leaving the opposite three sides are the same, no net mass change occurs within the box. More generally, flow leaving the opposite sides will be different because of change of velocity or density or both.

In the x direction, for example, the balance of flow in and out of the box is given by

$$\rho u r\delta\theta\delta r - \left(\rho u + \frac{\partial(\rho u)}{\partial x}\delta x\right) r\delta\theta\delta r \approx -\frac{\partial(\rho u)}{\partial x} r\delta\theta\delta r\delta x \tag{2.4.3}$$

the equality being approximate, again, because ρ and u may vary on that side of the box. The net result on the right represents mass flow *into* the box. Thus if the gradient on the right is positive, there will be net mass flow *out* of the box (because of the negative sign in front of that term). Similarly, the balance and net results in the directions of θ and r are respectively given by

$$\rho w\delta x\delta r - \left(\rho w + \frac{\partial(\rho w)}{\partial\theta}\delta\theta\right)\delta x\delta r \approx -\frac{\partial(\rho w)}{\partial\theta}\delta\theta\delta r\delta x \tag{2.4.4}$$

$$\begin{aligned}\rho v r\delta\theta\delta x &- \left(\rho v + \frac{\partial(\rho v)}{\partial r}\delta r\right)(r+\delta r)\delta\theta\delta x \\ &\approx -\rho v\delta r\delta\theta\delta x - \frac{\partial(\rho v)}{\partial r} r\delta\theta\delta r\delta x\end{aligned} \tag{2.4.5}$$

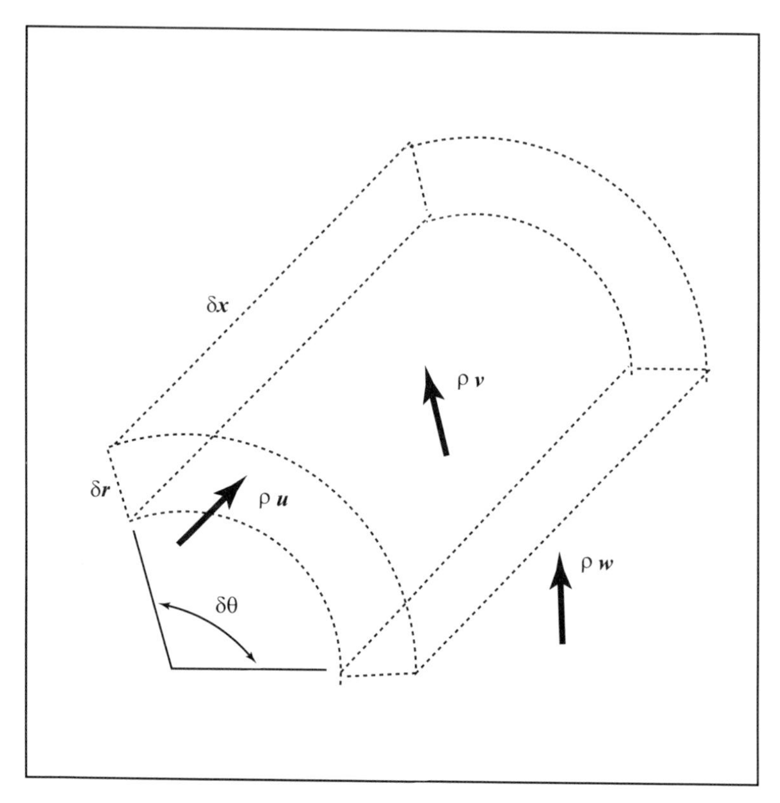

FIGURE 2.4.1. Conservation of mass. Rates of mass flow entering three sides of a box formed by coordinate surfaces are shown. If same flow rates *leave* the box through the opposite three sides, the net mass change within the box is zero. In general the rates will be different, however, because of changes in velocities or density. Conservation of mass requires that the difference between rates of mass flow into and out of the box be equal to the rate of change of mass of fluid within the box. As the box is shrunk to a point, this requirement becomes an equation of conservation of mass "at a point."

Conservation of mass requires that the total of these net changes in mass flow be equal to the rate of change of mass within the box, that is,

$$-\left\{\frac{\partial(\rho u)}{\partial x}+\frac{\partial(\rho v)}{\partial r}+\frac{\rho v}{r}+\frac{1}{r}\frac{\partial(\rho w)}{\partial\theta}\right\}r\delta\theta\delta r\delta x \approx \frac{\partial M}{\partial t}=\frac{\partial\rho}{\partial t}r\delta\theta\delta r\delta x \quad (2.4.6)$$

The volume of the box is a factor on both sides of the equation, which can thus be cancelled. Furthermore, as the volume of the box shrinks to a point, the density and velocities become the density ρ and velocities u, v, w

at that point. The equality thus becomes exact, and we obtain

$$\frac{\partial \rho}{\partial t} + \frac{\partial(\rho u)}{\partial x} + \frac{\partial(\rho v)}{\partial r} + \frac{\rho v}{r} + \frac{1}{r}\frac{\partial(\rho w)}{\partial \theta} = 0 \tag{2.4.7}$$

Finally, if the density does not vary from point to point within the flow field or with time, the equation takes on the simpler form

$$\frac{\partial u}{\partial x} + \frac{\partial v}{\partial r} + \frac{v}{r} + \frac{1}{r}\frac{\partial w}{\partial \theta} = 0 \tag{2.4.8}$$

This equation represents conservation of mass at a point in a flow field, although it is rarely referred to by this name. More commonly it is referred to as the "equation of continuity" because it is based on the assumption of continuity of the fluid body on the macroscopic scale, and because the equation in this form is only possible on that scale.

Since a point on the macroscopic scale represents a fluid element, the equation of continuity represents the law of conservation of mass applied to a fluid element. It is remarkable that the equation (Eq.2.4.7) does so without actually containing either the mass or volume of the fluid element. It contains only the density at a point. And when the density is constant, the equation in its simpler form (Eq.2.4.8) expresses the law of conservation of mass in terms of only the Eulerian velocity components, neither the mass nor density being involved.

Because conservation of mass is a fundamental physical law that must be satisfied in all circumstances, the equation of continuity is fundamental in the analysis of fluid flow problems and must be part of any set of equations governing real fluid flow.

2.5 Momentum Equations

Newton's law of motion applied to the momentum of a fluid element expresses the requirement that the mass times acceleration of the element be equal to the net of all forces acting on it. If m denotes the mass, a_x, a_r, a_θ denote the acceleration components in the directions of x, r, θ, respectively, and F_x, F_r, F_θ denote the corresponding components of the net force, then the equations of motion are of the form

$$\begin{aligned} ma_x &= F_x \\ ma_r &= F_r \\ ma_\theta &= F_\theta \end{aligned} \tag{2.5.1}$$

If each equation is divided by the volume v of the fluid element, the mass on the left-hand side of the equation becomes the density ρ at a point, and the forces on the right-hand side become *forces per unit volume.* Thus writing $f_x = F_x/v$, $f_r = F_r/v$, $f_\theta = F_\theta/v$, the equations of motion take the form

$$\begin{aligned} \rho a_x &= f_x \\ \rho a_r &= f_r \\ \rho a_\theta &= f_\theta \end{aligned} \tag{2.5.2}$$

As in the case of the equation of continuity, these equations express conditions at a point in a flow field. The conditions apply to the fluid element at that point, but by design neither the mass nor the volume of the element appear explicitly in the equations.

While the equations of motion are perfectly valid in this form, however, they are not readily usable because the forces on the right-hand side are not easily accessible. We shall see that the forces acting on a fluid element in a flow field result from a complex system of internal stresses that would be extremely difficult to access and measure in practice and are therefore highly unsuitable as variables in the equations of motion. The remaining challenge, therefore, is to express f_x, f_r, f_θ in terms of the previously chosen working variables, namely, the Eulerian velocity components u, v, w, and pressure p, which we do in the remainder of this chapter.

2.6 Forces on a Fluid Element

A fluid element in a flow field is typically surrounded by other elements that are in contact with it on all sides. There are two types of forces that can act on it: "boundary forces" exerted by neighboring elements and acting on the boundary of the element in question, and "body forces" exerted by some distant field of force and acting directly on the mass of the element.

Examples of body forces are those resulting from gravitational or magnetic fields. Since these act directly on the mass of the element, their form is simply the product of mass m and acceleration caused by the acting field of force, as in the case of an isolated body in a gravitational field, for example. In this book we shall not include these forces as they do not affect the basic understanding of pulsatile flow. Also, body forces when present in a flow field tend to be the main *driving* force, which provides the power required to maintain the flow, as in the case of flow in rivers. In physiological flow problems and in flow in tubes in general, by contrast, the main driving force is usually pressure, and as we shall see below, pressure is part of the complex system of boundary forces.

Pressure is indeed an important example of a boundary force, which is transmitted through the flow field from one fluid element to the next by direct action on the boundary of each element. Another example of a boundary force is the shear force created by a velocity gradient as seen in Section 1.7. It will be recalled, however, that the latter was a particularly simple case in which the force was caused by the gradient of only

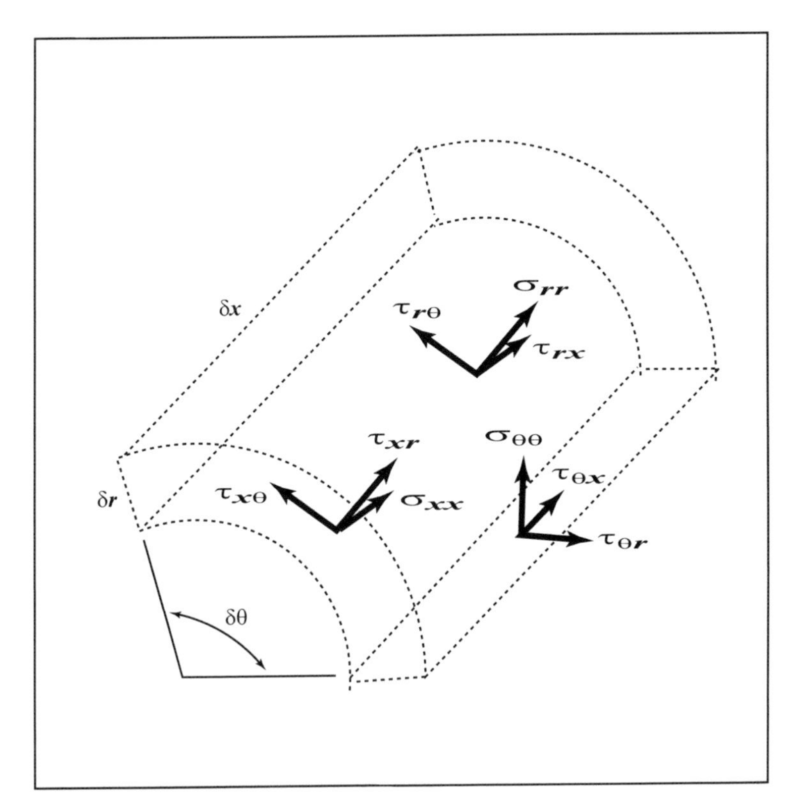

FIGURE 2.6.1. Stress tensor. Boundary stresses acting on three sides of a closed box formed by coordinate surfaces, a stress being force per area. On each side there is one stress acting normal to that side and two acting tangential to it. Normal stresses are denoted by σ's and tangential stresses by τ's, with the first subscript identifying the normal to the face on which the stress is acting and the second identifying the direction in which the stress itself is acting. Stresses acting on the opposite three sides may in general be different, thus producing a net force acting on the volume of fluid contained within the box. As the box is shrunk to a point, this force becomes the net boundary force acting "at a point," that is, acting on the fluid element at that point.

one velocity component and in only one direction. More generally, each of three velocity components may have gradients in each of three coordinate directions, thus giving rise to a complex system of shear forces acting on the boundary of a fluid element.

To study this system we again consider a closed box surrounding an element of fluid in a flow field, as we did in Section 2.4, but now examine the *forces* acting on each of the six sides of the box. In general there can be three forces acting on each side: one acting in a direction normal to the side and two acting tangentially to it. Forces on three sides of the box are therefore as shown in Fig.2.6.1. The notation is such that tangential forces are denoted by τ's and normal forces by σ's. In the subscript notation the first subscript indicates the direction of the normal to the surface on which the force is acting and the second indicates the direction in which the force itself is acting. The forces shown are in fact *forces per unit area* so that we shall henceforth refer to them more appropriately as *stresses.* The system of stresses acting on all sides is generally referred to as the *stress tensor*, and the individual stresses as its components.

The actual force arising from each stress component is thus the product of that stress and the area of the surface on which it is acting. For example, the area of the side normal to the x-axis is $r\delta\theta\delta r$ and the forces acting on it are approximately $\tau_{xr}r\delta\theta\delta r$ in the r direction, $\tau_{x\theta}r\delta\theta\delta r$ in the θ direction, and $\sigma_{xx}r\delta\theta\delta r$ in the x direction. The expressions are approximate because the τ's and σ's may vary over the surface of each side. Thus if all the forces on the three sides are grouped into those acting in the x, r, θ directions, respectively, we obtain, approximately again

$$\begin{aligned} &\sigma_{xx}r\delta\theta\delta r + \tau_{rx}r\delta\theta\delta x + \tau_{\theta x}\delta r\delta x \quad \text{in } x\text{-direction} \\ &\tau_{xr}r\delta\theta\delta r + \sigma_{rr}r\delta\theta\delta x + \tau_{\theta r}\delta r\delta x \quad \text{in } r\text{-direction} \\ &\tau_{x\theta}r\delta\theta\delta r + \tau_{r\theta}r\delta\theta\delta x + \sigma_{\theta\theta}\delta r\delta x \quad \text{in } \theta\text{-direction} \end{aligned} \tag{2.6.1}$$

The corresponding forces on the opposite three sides are generally different, the difference between each force and its counterpart on the opposite side being given by the incremental change in the force over the incremental distance between the two sides. For σ_{xx}, for example, the difference is given by

$$\begin{aligned} \left(\sigma_{xx}r\delta\theta\delta r + \frac{\partial(\sigma_{xx}r\delta\theta\delta r)}{\partial x}\delta x\right) \quad &- \quad \sigma_{xx}r\delta\theta\delta r \\ &\approx \quad \frac{\partial(\sigma_{xx}r\delta\theta\delta r)}{\partial x}\delta x \end{aligned} \tag{2.6.2}$$

The totals of all such differences, grouped into those acting in the x, r, θ directions, respectively, give the net boundary forces on the element. Thus the forces F_x, F_r, F_θ in Eq.2.5.1 are now given by

$$\begin{aligned} F_x &\approx \frac{\partial(\sigma_{xx}r\delta\theta\delta r)}{\partial x}\delta x + \frac{\partial(\tau_{rx}r\delta\theta\delta x)}{\partial r}\delta r + \frac{\partial(\tau_{\theta x}\delta r\delta x)}{\partial \theta}\delta\theta \\ F_r &\approx \frac{\partial(\tau_{xr}r\delta\theta\delta r)}{\partial x}\delta x + \frac{\partial(\sigma_{rr}r\delta\theta\delta x)}{\partial r}\delta r + \frac{\partial(\tau_{\theta r}\delta r\delta x)}{\partial \theta}\delta\theta \\ F_\theta &\approx \frac{\partial(\tau_{x\theta}r\delta\theta\delta r)}{\partial x}\delta x + \frac{\partial(\tau_{r\theta}r\delta\theta\delta x)}{\partial r}\delta r + \frac{\partial(\sigma_{\theta\theta}\delta r\delta x)}{\partial \theta}\delta\theta \end{aligned} \tag{2.6.3}$$

Upon division by the volume $V = r\delta\theta\delta r\delta x$, and as $V \to 0$, we find that the *forces per unit volume, at a point*, namely, f_x, f_r, f_θ in Eq.2.5.2, are given by

$$f_x = \frac{\partial \sigma_{xx}}{\partial x} + \frac{\partial \tau_{rx}}{\partial r} + \frac{\tau_{rx}}{r} + \frac{1}{r}\frac{\partial \tau_{\theta x}}{\partial \theta}$$

$$f_r = \frac{\partial \tau_{xr}}{\partial x} + \frac{\partial \sigma_{rr}}{\partial r} + \frac{\sigma_{rr}}{r} + \frac{1}{r}\frac{\partial \tau_{\theta r}}{\partial \theta}$$

$$f_\theta = \frac{\partial \tau_{x\theta}}{\partial x} + \frac{\partial \tau_{r\theta}}{\partial r} + \frac{\tau_{r\theta}}{r} + \frac{1}{r}\frac{\partial \sigma_{\theta\theta}}{\partial \theta} \tag{2.6.4}$$

The equalities are now no longer approximate because they are applied at a point and the values of all the functions involved are now their exact values at that point.

2.7 Extended Newtonian Relations: Constitutive Equations

The Newtonian relation discussed in Section 1.7 was based on only one velocity gradient and one shear stress τ. More generally, each of the three velocity components in a flow field may have a gradient in each of the three coordinate directions, and there are as many possible shear stress components on the boundary of a fluid element as we saw in the previous section. The relation between these velocity gradients and shear stresses for a given fluid is a defining property of the fluid known as its "constitutive equations." In the case of Newtonian fluids the relation is more complicated than that in Section 1.7, but it remains linear. *Linearity of the relation between shear stresses and velocity gradients is a defining characteristic of the constitutive equations of Newtonian fluids.*

Constitutive equations are generally based on a combination of theory and empirical data. For many common fluids the following equations are found to hold when velocity gradients are not large:

$$\begin{aligned}
\sigma_{xx} &= -p + 2\mu\left(\frac{\partial u}{\partial x}\right) \\
\sigma_{rr} &= -p + 2\mu\left(\frac{\partial v}{\partial r}\right) \\
\sigma_{\theta\theta} &= -p + 2\mu\left(\frac{1}{r}\frac{\partial w}{\partial \theta} + \frac{v}{r}\right) \qquad (2.7.1) \\
\tau_{xr} &= \tau_{rx} = \mu\left(\frac{\partial u}{\partial r} + \frac{\partial v}{\partial x}\right) \\
\tau_{x\theta} &= \tau_{\theta x} = \mu\left(\frac{\partial w}{\partial x} + \frac{1}{r}\frac{\partial u}{\partial \theta}\right)
\end{aligned}$$

$$\tau_{r\theta} = \tau_{\theta r} = \mu \left(\frac{\partial w}{\partial r} - \frac{w}{r} + \frac{1}{r}\frac{\partial v}{\partial \theta} \right) \tag{2.7.2}$$

Direct verification of constitutive equations is rarely possible. More commonly they are tested indirectly by using them as the basis of equations of fluid flow, as we do in the next section. The equations are subsequently solved for a given flow situation, and it is the *solutions* that are then compared directly with experiment. Such indirect verification, of course, does not test the assumptions on which the constitutive equations are based, and it is important to identify these at this point.

The central assumption is that of linearity of the relation between shear stress and velocity gradient. It is well recognized that this is only an approximation for small gradients, but its range of validity seems to be fairly wide. Solutions of equations of fluid flow based on the constitutive equations of Newtonian fluids have been applied to most common fluids and have been tested successfully against experiment for many years. Many problems in blood flow have been modeled successfully by these equations, even though the Newtonian character of blood can be questioned. In particular, the classical solutions for pulsatile flow that we use in this book have been based on the constitutive equations of Newtonian fluids.

The second important issue concerns the pressure p appearing in the constitutive equations. Pressure is strictly a thermodynamic property which is subject to the laws of thermodynamics applied to a body of fluid *at rest.* In that state the pressure represents a force that acts in the direction of the normal to the boundary of fluid elements and that is independent of the orientation of that boundary. The constitutive equations above are consistent with this state, as when the fluid is at rest ($u \equiv v \equiv w \equiv 0$) they reduce to

$$\sigma_{xx} = \sigma_{rr} = \sigma_{\theta\theta} = -p \tag{2.7.3}$$

Thus in this state the thermodynamic pressure is identified with the normal components of the stress tensor. The negative sign arises because normal stresses are defined to be positive in the direction of outward normal, while pressure in thermodynamics is defined to be positive in the direction of inward normal.

When fluid is in motion, however, the normal stresses are no longer equal to each other and the thermodynamic concept of pressure in fact no longer applies. What has been done to overcome this difficulty is to assume that the pressure in a moving fluid is equal to the *average* of the normal stresses, namely,

$$p = -\frac{(\sigma_{xx} + \sigma_{rr} + \sigma_{\theta\theta})}{3} \tag{2.7.4}$$

The constitutive equations are based on this assumption as can be readily verified by adding Eqs.2.7.1 and then using the equation of continuity

(Eq.2.4.8). The assumption inherent in this so-called "mechanical definition of pressure" was first made by Stokes, and it is on the basis of this assumption that the thermodynamic pressure p appears in the constitutive equations and subsequently in the equations of motion.

Finally, the constitutive equations assume that the stress tensor is symmetrical, namely,

$$\tau_{xr} = \tau_{rx}, \quad \tau_{r\theta} = \tau_{\theta r}, \quad \tau_{x\theta} = \tau_{\theta x} \tag{2.7.5}$$

and this again is recognized to be an approximation, but one which is found to work well under normal circumstances. Asymmetry of the stress tensor would give rise to forces that have spinning effects on fluid elements. The assumption of symmetry is based on the fact that such forces are found to be absent under normal flow conditions. An external field of force is required to produce them such as an electrostatic field, and in the absence of such fields the fluid does not support such asymmetry in the components of the stress tensor.

2.8 Navier–Stokes Equations

Having dealt with the acceleration of a fluid element and the system of forces acting on its boundary in the last two sections, it is now possible to complete the application of Newton's law of motion to a fluid element began in Section 2.5. Thus substituting for the acceleration terms from Eq.1.6.5 and for the boundary forces from Eqs.2.6.4, 2.7.1, 2.7.2, into Eqs.2.5.2, we obtain what is widely known as the Navier–Stokes equations in recognition of its first authors. The equations are sometimes also referred to as the momentum equations, as they actually govern the momentum of fluid elements:

$$\begin{aligned} &\rho\left(\frac{\partial u}{\partial t} + u\frac{\partial u}{\partial x} + v\frac{\partial u}{\partial r} + \frac{w}{r}\frac{\partial u}{\partial \theta}\right) + \frac{\partial p}{\partial x} \\ &= \mu\left(\frac{\partial^2 u}{\partial x^2} + \frac{\partial^2 u}{\partial r^2} + \frac{1}{r}\frac{\partial u}{\partial r} + \frac{1}{r^2}\frac{\partial^2 u}{\partial \theta^2}\right) \end{aligned} \tag{2.8.1}$$

$$\begin{aligned} &\rho\left(\frac{\partial v}{\partial t} + u\frac{\partial v}{\partial x} + v\frac{\partial v}{\partial r} + \frac{w}{r}\frac{\partial v}{\partial \theta} - \frac{w^2}{r}\right) + \frac{\partial p}{\partial r} \\ &= \mu\left(\frac{\partial^2 v}{\partial x^2} + \frac{\partial^2 v}{\partial r^2} + \frac{1}{r}\frac{\partial v}{\partial r} - \frac{v}{r^2} + \frac{1}{r^2}\frac{\partial^2 v}{\partial \theta^2} - \frac{2}{r^2}\frac{\partial w}{\partial \theta}\right) \end{aligned} \tag{2.8.2}$$

$$\begin{aligned} &\rho\left(\frac{\partial w}{\partial t} + u\frac{\partial w}{\partial x} + v\frac{\partial w}{\partial r} + \frac{w}{r}\frac{\partial w}{\partial \theta} + \frac{vw}{r}\right) + \frac{1}{r}\frac{\partial p}{\partial \theta} \\ &= \mu\left(\frac{\partial^2 w}{\partial x^2} + \frac{\partial^2 w}{\partial r^2} + \frac{1}{r}\frac{\partial w}{\partial r} - \frac{w}{r^2} + \frac{1}{r^2}\frac{\partial^2 w}{\partial \theta^2} + \frac{2}{r^2}\frac{\partial v}{\partial \theta}\right) \end{aligned} \tag{2.8.3}$$

These three equations, together with the equation of continuity (Eq.2.4.8) derived in Section 2.4, constitute a basic system of equations that govern a wide variety of fluid flow problems.

The Navier–Stokes equations and the equation of continuity govern the mass, momentum, and acceleration of fluid elements, and the forces acting on them, but none of these entities actually appear in the equations. Instead, they are all expressed in terms of the three Eulerian velocity components u, v, w, and pressure p. Solutions of the equations, therefore, provide information about the distribution of velocities and pressure in a flow field, which are generally the variables of interest for practical purposes.

It is important to recall that the equations of motion as they stand apply *at a point* in a flow field, that is, they apply to only an *element* of a fluid body. It is the solutions of these equations that finally provide information about the flow field and fluid body as a whole. In that solution process the equations are typically supplemented by the no-slip boundary condition, which applies at the interface between the fluid body and any solid boundaries. For flow in a tube, for example, the equations are supplemented by the condition of no-slip at the tube wall.

Finally, the equations in their full form are fairly general. In many cases they can be simplified by being specialized to a particular flow field, and the choice of coordinate system is an important element in that specialization. The choice of a cylindrical polar coordinate system which has already been made in this book is particularly suited for flow in a tube. It greatly facilitates the process of simplifying the governing equations as they are specialized to this particular flow, and we now turn to the details of this process.

2.9 Problems

1. Name the general laws of physics on which the equations governing fluid flow are based.

2. Explain what is meant by the equations of fluid flow being "point equations."

3. When density is constant, the law of conservation of mass as expressed by the equation of continuity (Eq.2.4.8) does not contain either mass or density. Explain how this is possible.

4. Physical laws on which the equations of fluid flow are based require the mass and identity of each element of a fluid body, because the laws apply to specific material objects. Explain how the final form of the equations (Eqs.2.8.1–3) contain neither the mass nor identity of individual elements.

5. Explain the difference between boundary forces and body forces that may act on a fluid element in a flow field. Give examples of each and discuss these in the context of flow in a tube.
6. Identify the component of the stress tensor that acts at the tube wall to the effect of opposing the flow in a tube. State the conditions under which this stress (τ) is given by

$$\tau = \mu \frac{\partial u}{\partial r}$$

where u is velocity component in the axial direction, r is radial coordinate, and μ is viscosity of the fluid.
7. In the first Navier–Stokes equation (Eq.2.8.1) identify the form "*force = mass × acceleration*" and hence the terms which represent its three elements in the equation.
8. Write down the Navier–Stokes equations and equation of continuity in rectangular Cartesian coordinates x, y, z and corresponding velocity components u, v, w, hence showing the difference between these and the corresponding equations in cylindrical polar coordinates (Eqs.2.8.1–3, 2.4.8).

2.10 References and Further Reading

1. Batchelor GK, 1967. An Introduction to Fluid Dynamics. Cambridge University Press, Cambridge.
2. Curle N, Davies HJ, 1968. Modern Fluid Dynamics: I. Incompressible Flow. Van Nostrand, Princeton, New Jersey.
3. Duncan WJ, Thom AS, Young AD, 1970. Mechanics of Fluids. Edward Arnold, London.
4. McCormack PD, Crane L, 1973. Physical Fluid Dynamics. Academic Press, New York.
5. Munson BR, Young DF, Okiishi TH, 1990. Fundamentals of Fluid Mechanics. John Wiley, New York.
6. Panton RL, 1984. Incompressible Flow. John Wiley, New York.
7. Rosenhead L, 1963. Laminar Boundary Layers. Oxford University Press, Oxford.
8. Schlichting H, 1979. Boundary Layer Theory. McGraw-Hill, New York.
9. Tritton DJ, 1988. Physical Fluid Dynamics. Clarendon Press, Oxford.

3
Steady Flow in Tubes

3.1 Introduction

When flow enters a tube, the no-slip boundary condition on the tube wall arrests fluid elements in contact with the wall while elements along the axis of the tube charge ahead, less influenced by that condition. Because viscosity of the fluid does not allow a step change in velocity to occur anywhere in the flow field, a smooth velocity profile develops to join the faster moving fluid along the axis of the tube with the stationary fluid at the tube wall.

At the entrance to the tube, only fluid actually in contact with the tube wall is affected by the no-slip boundary condition, thus the velocity profile consists of a straight line representing the bulk of the fluid moving uniformly down the tube, then dropping rapidly but smoothly to zero near the wall. Further and further downstream, however, the arrested layer of fluid at the tube wall begins to slow down the layer of fluid in contact with it on the other side, and this effect propagates further and further away from the wall, in effect owing to loss of energy by viscous dissipation at the tube wall. Thus the region of influence of the no-slip boundary condition grows further and further away from the wall, with the result that the velocity profile becomes more and more rounded.

When the wall effect engulfs the entire cross section of the tube, the velocity profile reaches equilibrium and does not change further. This part of the flow field is referred to as the "fully developed" region, while that preceeding it is referred to as the "entry flow" region (Fig.3.1.1).

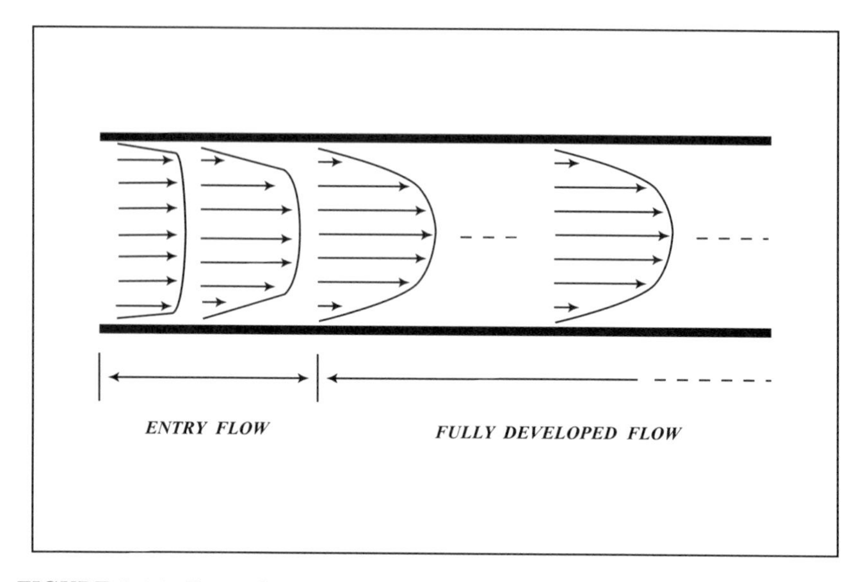

FIGURE 3.1.1. Entry flow in a tube. When flow enters a tube, only fluid near the tube wall or in contact with it is influenced by the no-slip boundary condition there. As flow moves further downstream, however, this region of influence grows, leading ultimately to a more rounded velocity profile. Flow in this region is called "fully developed" in the sense that it has reached its ultimate form, while developing flow in the preceding region is called "entry flow."

The rate of energy expenditure required to overcome viscous losses at the tube wall and inertial effects near the tube axis can be provided by a pressure difference between the two ends of the tube, or by a gravitational force acting as a body force directly on the mass of fluid elements as in the case of flow in an inclined tube. In physiological applications the first of these is more pertinent, as the physics of blood flow is governed far more by the pressure difference created by the pumping action of the heart than by the effects of gravity.

3.2 Simplified Equations

In its most general form, flow in a tube requires the full Navier–Stokes equations and the equation of continuity, but if it can be assumed that the cross section of the tube is circular and the tube is straight and sufficiently long, and if attention is focused on only the fully developed region of the flow, these equations can be simplified considerably. The classical solutions for pulsatile flow are based on a highly simplified form of the equations, and in this section we present the assumptions on which the simplifications are based and the steps in the simplifying process.

If the tube is straight and has a circular cross section, and in the absence of any external forces that would cause flow rotation, the flow field will be symmetrical about the longitudinal axis of the tube to the effect that the angular component of velocity and all derivatives in the angular direction are then zero, that is,

$$w \equiv \frac{\partial w}{\partial \theta} \equiv \frac{\partial v}{\partial \theta} \equiv \frac{\partial u}{\partial \theta} \equiv \frac{\partial p}{\partial \theta} \equiv 0 \tag{3.2.1}$$

All terms in the w equation of motion (Eq.2.8.3) are then identically zero and the other two equations, together with the equation of continuity (Eqs.2.8.1, 2.4.8) simplify to

$$\rho\left(\frac{\partial u}{\partial t}+u\frac{\partial u}{\partial x}+v\frac{\partial u}{\partial r}\right)+\frac{\partial p}{\partial x} = \mu\left(\frac{\partial^2 u}{\partial x^2}+\frac{\partial^2 u}{\partial r^2}+\frac{1}{r}\frac{\partial u}{\partial r}\right) \tag{3.2.2}$$

$$\rho\left(\frac{\partial v}{\partial t}+u\frac{\partial v}{\partial x}+v\frac{\partial v}{\partial r}\right)+\frac{\partial p}{\partial r} = \mu\left(\frac{\partial^2 v}{\partial x^2}+\frac{\partial^2 v}{\partial r^2}+\frac{1}{r}\frac{\partial v}{\partial r}-\frac{v}{r^2}\right) \tag{3.2.3}$$

$$\frac{\partial u}{\partial x}+\frac{\partial v}{\partial r}+\frac{v}{r}=0 \tag{3.2.4}$$

If these equations are now further restricted to only the fully developed region of the flow, where by definition

$$\frac{\partial u}{\partial x} \equiv \frac{\partial v}{\partial x} \equiv 0 \tag{3.2.5}$$

then the equation of continuity reduces to

$$\frac{\partial v}{\partial r}+\frac{v}{r}=\frac{1}{r}\frac{\partial (rv)}{\partial r}=0 \tag{3.2.6}$$

which can be integrated to give $rv = constant$. And because v must be zero at the tube wall ($r = a$), this result implies that the radial component of velocity must be identically zero, that is,

$$v \equiv 0 \tag{3.2.7}$$

As a result of this condition and that in Eq.3.2.5, the equation of continuity (Eq.3.2.4) is now satisfied identically. For the same reasons all the velocity terms in the v equation of motion (Eq.3.2.3) are now zero, which therefore implies

$$\frac{\partial p}{\partial r} \equiv 0 \tag{3.2.8}$$

In the remaining equation of motion (Eq.3.2.2), terms containing velocity gradients in x are zero because of Eq.3.2.5 and the term containing v is zero because of Eq.3.2.7, thus the equation reduces to

$$\rho \frac{\partial u}{\partial t} + \frac{\partial p}{\partial x} = \mu \left(\frac{\partial^2 u}{\partial r^2} + \frac{1}{r}\frac{\partial u}{\partial r} \right) \tag{3.2.9}$$

This is the highly simplified form of the governing equations on which the classical solutions for fully developed steady and pulsatile flow are based. As a result of the simplifying assumptions and their consequences in Eqs.3.2.5 and 3.2.8, the velocity u in Eq.3.2.9 is now a function of r and t only, while the pressure p is a function of x and t only, that is,

$$u = u(r,t), \quad p = p(x,t) \tag{3.2.10}$$

It is important to note here that the only way in which the pressure can vary along the tube but not the velocity is for the tube to be *rigid.* If the tube is not rigid, a local change in pressure causes a local change in the cross section of the tube and hence a change in velocity.

3.3 Steady–State Solution: Poiseuille Flow

If the pressure difference driving the flow in a tube is not a function of time, the velocity field within the tube will also be independent of time and a steady state will prevail where Eqs.3.2.10 become

$$u = u_s(r), \quad p = p_s(x) \tag{3.3.1}$$

where the velocity and pressure are being identified with a subscript s as a reference to this steady state. Again we recall that the only way in which these two conditions can be satisfied simultaneously is for the tube to be *rigid.* The time derivative term in Eq.3.2.9 is now zero and the remaining derivatives become ordinary derivatives; thus, the equation reduces to

$$\frac{dp_s}{dx} = \mu \left(\frac{d^2 u_s}{dr^2} + \frac{1}{r}\frac{du_s}{dr} \right) \tag{3.3.2}$$

An important feature of this flow is that its governing equation is independent of the density ρ. The reason for this is that acceleration terms are now zero. The remaining terms in the equation represent a balance of forces between only the driving pressure and the viscous resistance of the fluid.

In Eq.3.3.2 the term on the left-hand side is a function of x only while the terms on the right-hand side are functions of r only. The only way in which the equation can be satisfied in general, therefore, is by having both sides equal a constant, the same constant, say k_s, thus

$$\frac{dp_s}{dx} = k_s \tag{3.3.3}$$

$$\mu \left(\frac{d^2 u_s}{dr^2} + \frac{1}{r} \frac{du_s}{dr} \right) = k_s \tag{3.3.4}$$

Solving the first equation gives

$$p_s(x) = p_s(0) + k_s x \tag{3.3.5}$$

where x is axial distance from the tube entrance. If the exit from the tube is at $x = l$, then the constant k_s is given by

$$k_s = \frac{p_s(l) - p_s(0)}{l} \tag{3.3.6}$$

Solving the second equation gives

$$u_s(r) = \frac{k_s}{4\mu} r^2 + A \ln r + B \tag{3.3.7}$$

where A, B are constants of integration. The two boundary conditions for evaluating these are no-slip at the tube wall and finite velocity at the center, that is

$$u_s(a) = 0, \quad |u_s(0)| < \infty \tag{3.3.8}$$

which give

$$A = 0, \quad B = -\frac{k_s a^2}{4\mu} \tag{3.3.9}$$

With these values the solution becomes finally

$$u_s = \frac{k_s}{4\mu}(r^2 - a^2) \tag{3.3.10}$$

This is the classical solution for steady flow in a tube, usually referred to as *Poiseuille flow* after its first author [1,2].

3.4 Properties of Poiseuille Flow

The velocity profile obtained in the previous section has the characteristic parabolic form commonly associated with flow in a tube. It indicates that maximum velocity occurs on the axis of the tube ($r = 0$) as expected on physical grounds, and zero velocity occurs on the tube wall ($r = a$) as dictated by the no-slip boundary condition, that is,

$$\hat{u}_s = u_s(0) = \frac{-k_s a^2}{4\mu}, \quad u_s(a) = 0 \tag{3.4.1}$$

The minus sign in the first equation indicates that the direction of velocity is opposite to that of the pressure gradient k_s, that is, velocity is positive in the direction of negative pressure gradient. As anticipated, maximum velocity on the axis of the tube and zero velocity at the tube wall are

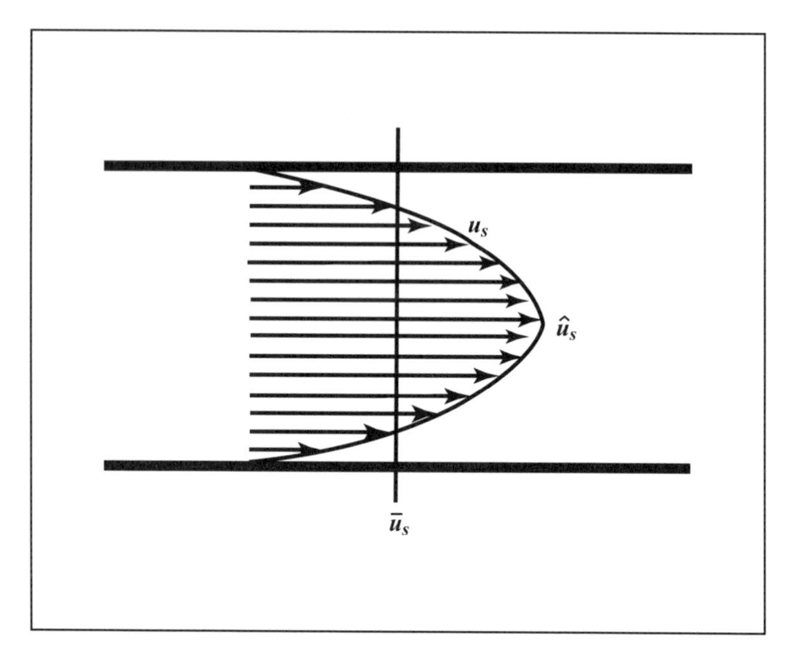

FIGURE 3.4.1. Velocity profile in steady fully developed Poiseuille flow (u_s). The profile has a parabolic form and the average velocity ($\overline{u}_s$) is one half the maximum velocity ($\hat{u}_s$).

joined by a *smooth* profile, with no step change at any point in between. It is convenient to nondimensionalize the velocity profile in terms of the maximum velocity, to get

$$\frac{u_s(r)}{\hat{u}_s} = 1 - \left(\frac{r}{a}\right)^2 \tag{3.4.2}$$

Volumetric flow rate q_s through the tube is obtained by integrating over a cross section of the tube

$$q_s = \int_0^a u_s 2\pi r dr = \frac{-k_s \pi a^4}{8\mu} \tag{3.4.3}$$

Again, the minus sign indicates that flow rate and pressure gradient have opposite signs. That is, negative pressure gradient produces flow in the positive x direction. The average velocity $\overline{u}_s$ is given by

$$\overline{u}_s = \frac{q_s}{\pi a^2} = \frac{-k_s a^2}{8\mu} \tag{3.4.4}$$

which is seen to be one half the maximum velocity on the tube axis (Eq.3.4.1, Fig.3.4.1).

The radial and angular velocity components v, w are zero in Poiseuille flow, and the axial velocity component is a function of r only. Because of this, only one of the shear stress components defined in Eq.2.7.2 is nonzero, and it has the simpler form

$$\tau_{xr} = \tau_{rx} = \mu \frac{du}{dr} \tag{3.4.5}$$

As seen in Fig.2.6.1 the first of these is in planes perpendicular to x and acts in the direction of r, while the second is in cylindrical surfaces parallel to the tube wall and acts in the directions of x. The second is of particular interest because it produces the resistance to flow at the tube wall that we shall denote by $-\tau_s$, thus

$$\tau_s = -\tau_{rx}(a) = -\mu \left(\frac{du_s}{dr} \right)_{r=a} = \frac{-k_s a}{2} \tag{3.4.6}$$

the sign difference between τ_s and τ_{rx} is there because τ_s is here taken to represent the shear stress exerted by the fluid on the tube wall. Substituting for k_s from Eq.3.4.3, finally

$$\tau_s = \frac{4\mu q_s}{\pi a^3} \tag{3.4.7}$$

The rate of energy expenditure required to overcome the resistance to flow at the tube wall, that is, the pumping power H_s required to maintain the flow, is given by the product of the total force of resistance and the average flow velocity. The total force in turn is the product of the shear stress and the surface area on which it is acting, which is the surface area of the tube wall, that is,

$$H_s = \tau_s \times 2\pi a l \times \overline{u}_s \tag{3.4.8}$$

where l is the tube length. If this resistance is overcome by a pressure difference $\triangle p_s$, where (Fig.3.4.2)

$$\triangle p_s = p_s(l) - p_s(0) = k_s l \tag{3.4.9}$$

then the power is also equal to the product of the total driving force and the average velocity through the tube, that is,

$$H_s = - \triangle p_s \times \pi a^2 \times \overline{u}_s \tag{3.4.10}$$

Equality of these two forms is evident from Eqs.3.4.6 and 3.4.9.

The pressure difference can be expressed in terms of the flow rate, using Eqs.3.4.3 and 3.4.9

$$- \triangle p_s = \left(\frac{8\mu l}{\pi a^4} \right) q_s \tag{3.4.11}$$

Again, the minus sign indicates that $\triangle p_s$ and q_s have opposite signs, as explained earlier. This relation is analogous to that of the flow of current in an electric conductor, namely,

$$E = RI \tag{3.4.12}$$

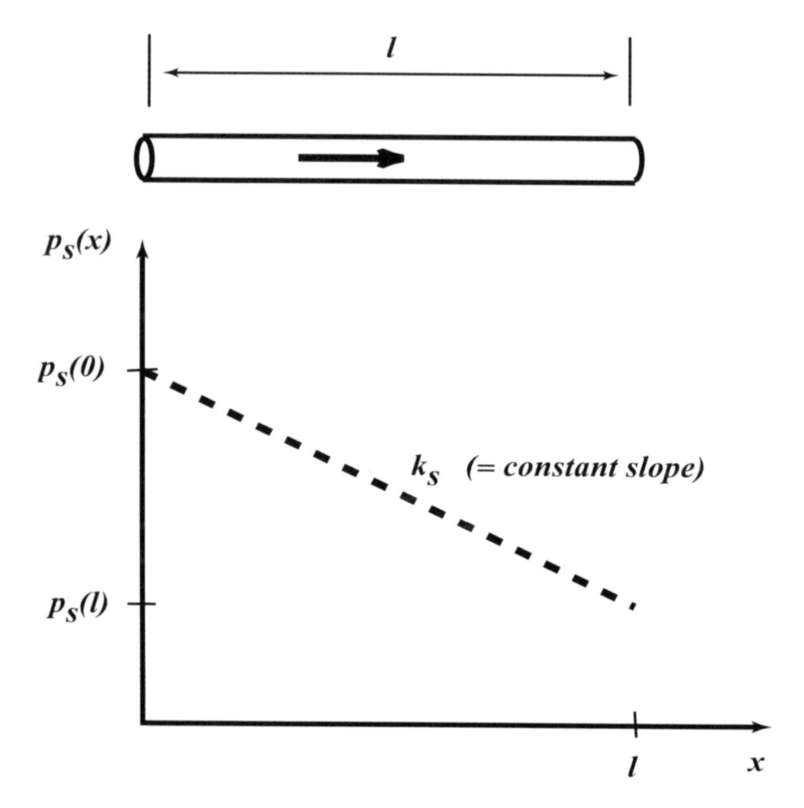

FIGURE 3.4.2. Pressure distribution in steady fully developed Poiseuille flow. Pressure $p_s(x)$ decreases linearly along the tube. Pressure gradient k_s is constant, and pressure difference $p_s(0) - p_s(l)$ is a measure of the "force" required to maintain the flow against the shear force at the tube wall.

where E is the potential difference, R is the resistance and I is the current. In the context of this analogy, the pressure difference in Eq.3.4.11 is identified with E, the bracketed term is identified with R, and the volumentric flow rate is identified with I. Thus in Poiseuille flow *resistance to flow is inversely proportional to the fourth power of the radius.*

Finally, the expression for the pumping power in Eq.3.4.10 can also be put in the simpler form

$$H_s = - \triangle p_s \times q_s \tag{3.4.13}$$

which again is consistent with the electric analogy in which power equals the product of potential difference and current. Substituting for the pressure difference from Eq.3.4.11, the power can finally be put in the form

$$H_s = -lk_s q_s = \left(\frac{8\mu l}{\pi a^4}\right) q_s^2 \tag{3.4.14}$$

from which it is seen that for a fixed tube radius pumping power is *directly proportional to the second power of the flow rate*, and for a fixed flow rate it is *inversely proportional to the fourth power of the radius.*

3.5 Balance of Energy Expenditure

The equation governing steady flow in a tube, Eq.3.3.4,

$$\mu \left(\frac{d^2 u_s}{dr^2} + \frac{1}{r} \frac{du_s}{dr} \right) = k_s$$

expresses a balance between the forces affecting the flow, specifically the driving pressure force on the right-hand side and the retarding viscous force on the left-hand side. At any point in *time* during flow in a tube, these two forces are associated with rates of energy expenditure. The pressure force is associated with the rate of energy expenditure required to drive the flow, or the "pumping power," while the viscous force is associated with the rate of energy dissipation by viscosity. In this section we examine these two rates of energy expenditure and the way in which their balance is based on the equation governing the flow. While in the present case both rates are constant and the balance between them is one of simple equality, this simple case serves as a useful foundation for the same exercise in pulsatile flow that we consider in the next chapter.

As it stands, the equation of motion above actually represents not forces but *forces per unit volume.* This, we recall, is a basic characteristic of the more general Navier–Stokes equations from which the above equation is obtained. In order to consider energy expenditures, therefore, the equation must first be multiplied by some volume of fluid so that its terms will actually represent forces, then determine the rate at which these forces are doing work. It should be recalled that work or energy is produced by force times distance, while power or rate of energy expenditure is produced by force times distance per time, or force times velocity.

It is therefore necessary to consider a small volume of fluid which is moving with the same velocity. A convenient choice is that of a thin cylindrical shell of radius r, thickness dr, and length l. The volume of fluid comprising the shell is then $2\pi r l dr$ and the velocity with which its elements are moving is simply the Poiseuille flow velocity at radial position r, namely, $u_s(r)$. If the equation of motion above is multiplied by the volume of this shell and by the velocity with which it is moving, the result is an equation representing the balance of energy expenditure associated with this volume of fluid, namely,

$$\mu \left(\frac{d^2 u_s}{dr^2} + \frac{1}{r} \frac{du_s}{dr} \right) \times 2\pi r l u_s dr = k_s \times 2\pi r l u_s dr \tag{3.5.1}$$

Furthermore, if each side of this equation is now integrated over a cross section of the tube, that is, from $r = 0$ to $r = a$, the same balance will be established for fluid filling the entire volume of a tube of radius a and length l.

The resulting integral on the right-hand side is then

$$\begin{aligned} \int_0^a k_s \times 2\pi r l u_s dr &= k_s l \int_0^a 2\pi r u_s dr \\ &= k_s l q_s \end{aligned} \tag{3.5.2}$$

where q_s is the flow rate through the tube as determined by Eq.3.4.3. The integral on the left-hand side is

$$\begin{aligned} &\int_0^a \mu \left(\frac{d^2 u_s}{dr^2} + \frac{1}{r}\frac{du_s}{dr} \right) \times 2\pi r l u_s dr \\ &= 2\pi\mu l \int_0^a u_s \frac{d}{dr}\left(r\frac{du_s}{dr} \right) dr \\ &= 2\pi\mu l \int_{r=0}^{r=a} u_s d\left(r\frac{du_s}{dr} \right) \\ &= 2\pi\mu l \left\{ \left| u_s r\frac{du_s}{dr} \right|_{r=0}^{r=a} - \int_{r=0}^{r=a} r\frac{du_s}{dr} du_s \right\} \\ &= -2\pi\mu l \int_0^a r \left(\frac{du_s}{dr} \right)^2 dr \end{aligned} \tag{3.5.3}$$

Equality of the two results in Eqs.3.5.2,3 thus establishes the balance of energy expenditure within the tube, namely,

$$2\pi\mu l \int_0^a r \left(\frac{du_s}{dr} \right)^2 dr = -k_s l q_s \tag{3.5.4}$$

The term on the right-hand side will be recognized as the pumping power required to drive the flow, as determined previously in Eq.3.4.14. The term on the left-hand side represents the rate of energy dissipation by viscosity and it should equal that found previously in Eq.3.4.8. To verify this we substitute for u_s in Eq.3.5.4 from Eq.3.3.10 to get

$$\begin{aligned} 2\pi\mu l \int_0^a r \left(\frac{du_s}{dr} \right)^2 dr &= 2\pi\mu l \left(\frac{k_s}{2\mu} \right)^2 \int_0^a r^3 dr \\ &= 2\pi\mu l \left(\frac{k_s}{2\mu} \right)^2 \left(\frac{a^4}{4} \right) \\ &= 2\pi a l \left(\frac{k_s a^2}{8\mu} \right) \left(\frac{k_s a}{2} \right) \end{aligned}$$

$$= 2\pi al(\overline{u}_s)(\tau_s) \tag{3.5.5}$$

where Eqs.3.4.4,6 have been used for $\overline{u}_s$ and τ_s in the last step. This final result is now seen to be identical with that obtained earlier for the power required to drive the flow (Eq.3.4.8). Thus the term on the left-hand side of Eq.3.5.4 is indeed equal to the power required to drive the flow which in turn is equal to the rate of viscous dissipation at the tube wall.

3.6 Cube Law

While flow in the cardiovascular system is pulsatile, some basic design features of the arterial tree can be examined fairly adequately in terms of steady Poiseuille flow. Results of these considerations and results of Poiseuille flow in general serve later as important foundations for similar considerations in pulsatile flow.

If the flow in a blood vessel is assumed to be the same as fully developed Poiseuille flow in a tube, a fundamental question concerning the structure of arterial trees can be addressed, namely, that of the relation between the caliber of a vessel and the flow rate that the vessel is destined to convey.

From results of the previous section we note that the pumping power H_s required to maintain a flow rate q through a tube of radius a is inversely proportional to the fourth power of the radius, which suggests that from the point of view of fluid dynamics the vessel radius should be as large as possible (Fig.3.6.1). From the biological point of view, however, a vessel of larger radius requires a larger volume of blood to fill, and the latter in turn requires a higher rate of metabolic energy to maintain. Assuming that this metabolic power is proportional to the volume of the vessel, then for a given length of vessel it will be proportional to a^2 while the pumping power is proportional to a^{-4}.

Thus a simple optimality problem is established, which was first presented by Cecil D. Murray in 1926 [3,4], namely that the total power H required for both fluid dynamic and biological purposes is given by

$$H = \frac{A}{a^4} + Ba^2 \tag{3.6.1}$$

where B is a positive constant representing the metabolic rate of energy required to maintain the volume of blood, and from Eq.3.4.14

$$A = \frac{8\mu l q_s^2}{\pi} \tag{3.6.2}$$

A minimum value of H occurs when

$$\frac{dH}{da} = \frac{-4A}{a^5} + 2Ba = 0, \quad \frac{d^2H}{da^2} = \frac{20A}{a^6} + 2B > 0 \tag{3.6.3}$$

Because both A and B are positive, the inequality is satisfied and the first equation gives

$$a^6 = \frac{2A}{B} = \frac{2}{B}\frac{8\mu l}{\pi}q_s^2 \tag{3.6.4}$$

Because B is a constant that does not depend on the flow rate, it follows from Eq.3.6.4 that the required condition of minimum power occurs when

$$q_s \propto a^3 \tag{3.6.5}$$

which is usually referred to as the "cube law," or as "Murray's law" after its first author.

The cube law has been used for many years and has shown considerable presence in the cardiovascular system [5–14]. It is important to recall that it rests on two important assumptions: (i) that the flow under consideration is steady fully developed Poiseuille flow, and (ii) that the optimality criterion being used is that of minimizing the total rate of energy expenditure for dynamic and metabolic purposes.

Other laws have been considered by a number of authors to address observed departures or scatter away from the cube law [15–17]. It has been found, for example, that in the aorta and its first generation of major

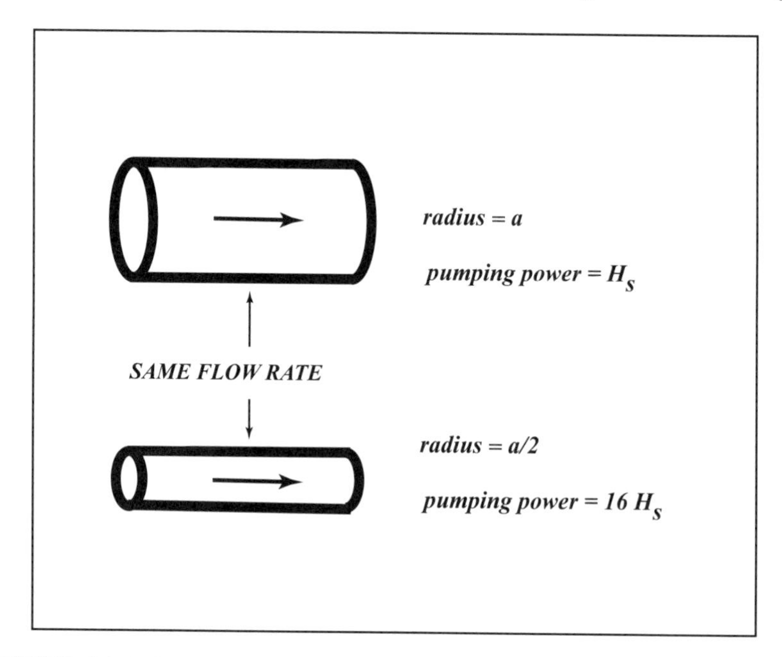

FIGURE 3.6.1. Pumping power required to maintain steady fully developed Poiseuille flow. If the radius of a tube is halved, the pumping power required to drive the same flow rate through it increases by a factor of 16 (1600%). Conversely, if the radius of a tube is doubled, only about 6% of the power is required to drive the same flow rate through it.

branches, a "square law" in which the flow rate is proportional to a^2 may be more appropriate than the cube law. At more peripheral regions of the arterial tree, however, measured data suggest that the cube law eventually prevails, even if with considerable scatter.

Because in Poiseuille flow the shear stress on the tube wall is proportional to the ratio q/a^3 (Eq.3.4.7), the cube law is consistent with constant shear rate in the vascular system. That is, at higher levels of the arterial tree, as the diameters of vessel segments become smaller, flow rates within them also become smaller but in accordance with the cube law, leaving the shear stress on vessel walls unchanged. Because the range of diameters in the arterial tree extends over three to four orders of magnitude, any departure from this precarious design would lead to a range of shear stress of many orders of magnitude. If the flow rate varies in accordance with a square law ($q \propto a^2$) or a quartic law ($q \propto a^4$), for example, the shear stress on the vessel walls would be proportional to $1/a$ or to a, respectively. Thus as a varies by three to four orders of magnitude, so will the shear stress, and in the case of the square law the higher shear will occur in the smaller vessels. Because the lumen of vessels of all sizes are lined with essentially the same type of endothelial tissue, these scenarios seem unlikely on theoretical grounds. Thus the form of the shear stress in Poiseuille flow provides a strong theoretical support for the cube law, on grounds that are quite different from those on which the law was originally based.

3.7 Arterial Bifurcation

The principal structural unit of an arterial tree is a "bifurcation," whereby a parent vessel segment divides into two daughter segments. Data from the cardiovascular system have shown that division into more than two daughters is rare [7–12]. The three vessel segments at an arterial bifurcation usually lie in the same plane, but a change in the orientation of this plane from one bifurcation to the next produces highly three dimensional-tree structures [18,19]. Also, wide variations in the degree of asymmetry of arterial bifurcations, that is, wide variation in the degree to which the two daughter vessels are of unequal caliber, makes it possible for arterial trees to be highly nonuniform and highly nonsymmetrical, as required by the particular territory which they serve [36].

If the radii of the parent and daughter vessel segments at an arterial bifurcation are denoted by a_0, a_1, a_2, and their diameters are denoted by d_0, d_1, d_2 (Fig.3.7.1), and if we use the convention of always taking $a_1 \geq a_2$, then a useful *bifurcation index* can be defined by

$$\alpha = \frac{a_2}{a_1} \tag{3.7.1}$$

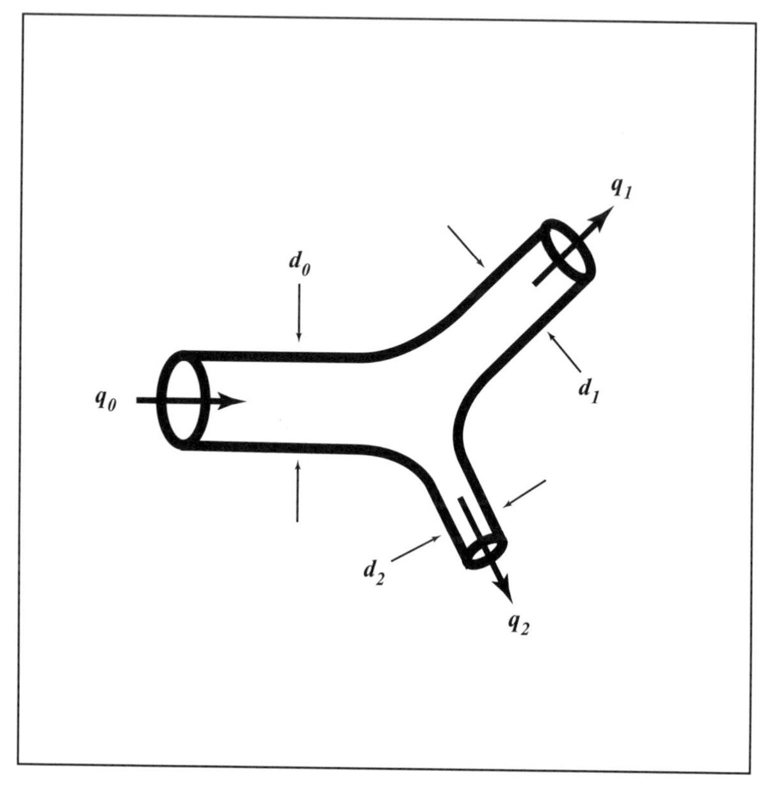

FIGURE 3.7.1. Arterial bifurcation, the basic structural unit of an arterial tree. A parent vessel of diameter d_0 divides into two daughter vessels of diameters d_1, d_2. Conservation of mass requires that flow rate q_0 in the parent vessel must equal the sum of flow rates in the daughter vessels, $q_1 + q_2$. Combined with the cube law, this provides an important "optimum" relation between the three diameters involved, namely, $d_0^3 = d_1^3 + d_2^3$.

and its value ranges conveniently between 0 and 1.0. A highly nonsymmetrical bifurcation is one for which the value of α is near zero, while a "symmetrical" bifurcation is one for which $\alpha = 1.0$.

Another important measure at an arterial bifurcation is the *area ratio*

$$\beta = \frac{a_1^2 + a_2^2}{a_0^2} \tag{3.7.2}$$

which is the ratio of the combined cross-sectional area of the two daughters over that of the parent vessel. Values of β greater than 1.0 produce expansion in the total cross-sectional area available to flow as it progresses from one level of the tree to the next.

Conservation of mass requires that at an arterial bifurcation flow rate in the parent vessel must equal the sum of those in the two daughter vessels;

thus, if these rates are denoted by q_0, q_1, q_2, respectively, then

$$q_0 = q_1 + q_2 \tag{3.7.3}$$

If the cube law is assumed to hold, this relation between flow rates becomes one between the radii of the three vessels at an arterial bifurcation, namely,

$$a_0^3 = a_1^3 + a_2^3 \tag{3.7.4}$$

In terms of the bifurcation index α, this relation yields

$$\frac{a_1}{a_0} = \frac{1}{(1+\alpha^3)^{1/3}}$$

$$\frac{a_2}{a_0} = \frac{\alpha}{(1+\alpha^3)^{1/3}} \tag{3.7.5}$$

and substituting these into the expression for the area ratio β (Eq.3.7.2), gives

$$\beta = \frac{1+\alpha^2}{(1+\alpha^3)^{2/3}} \tag{3.7.6}$$

Because these results are based on the cube law, they represent "optimum" values of the relative radii and area ratio at an arterial bifurcation, the optimality criterion being the same as that on which the cube law is based. It must be recalled that the cube law is also based on the simplifying assumptions associated with Poiseuille flow, thus the above results are equally based on these assumptions. Despite these limitations, however, the results have found considerable support from measured data [5–12]. For a symmetrical bifurcation ($\alpha = 1.0$), in particular, the results in Eqs.3.7.5,6 reduce to

$$\frac{a_1}{a_0} = \frac{a_2}{a_0} = 2^{-1/3} \approx 0.7937 \tag{3.7.7}$$

$$\beta = 2^{1/3} \approx 1.2599 \tag{3.7.8}$$

These values are useful for approximate calculations of the global properties of theoretical tree structures, which are usually based on the assumption that the tree consists of symmetrical bifurcations.

3.8 Arterial Tree

Arterial trees in the cardiovascular system (Fig.1.1.1) serve the purpose of bringing blood flow to many millions of tissue cells, *individually*. An open tree structure whereby a parent vessel undergoes a bifurcation, then each of the two daughter vessels in turn undergoes a bifurcation, and so on, makes it possible to bring that flow from a single source and distribute it to many destinations, without duplication. The branching structure of the tree allows the "sharing" of blood vessels so that the individual flow to each tissue cell does not have to be carried individually from source to destination. *The tree structure eliminates the need to run tubes in parallel.* Saving occurs at each bifurcation and is multiplied many times over by the large number of bifurcations that comprise the tree.

At an arterial bifurcation, blood flow from point A at the entrance of the parent vessel is destined to reach points B and C at the exits from the two daughter vessels (Fig.3.8.1). Instead of running two tubes in parallel from A to B and from A to C, the bifurcation makes it possible to share the flow from A to some junction point J before dividing into two separate tubes. The saving is approximately that of running only one tube from A to J instead of two.

In Poiseuille flow the power H required to drive a flow rate q through a tube of radius a is proportional to q^2a^{-4} (Eq.3.4.14), and if the cube law is assumed to hold (Eq.3.6.5), this gives $H \propto a^2$. Thus if the powers required to drive the flow in tubes of radii a_0, a_1, a_2 are denoted by H_0, H_1, H_2, then the fractional difference between running two tubes of radii a_1, a_2 from A to J or only one tube of radius a_0, is given by, in nondimensional form,

$$\frac{H_1 + H_2 - H_0}{H_0} = \frac{a_1^2 + a_2^2 - a_0^2}{a_0^2} = \beta - 1 \tag{3.8.1}$$

Because the value of β is generally higher than 1.0, the result is positive. If the bifurcation is symmetrical and we take $\beta \approx 1.26$ as determined by the cube law (Eq.3.7.8), a saving of 26% is realized by running a single tube instead of two.

This calculation is only approximate, however, because a straight line from A to B would be shorter than one from A to J plus one from J to B, and similarly for a line from A to C. In fact to account for these differences an optimality problem must be solved for the position of the junction point J, which determines the branching angles θ_1, θ_2 which the two daughter vessels should optimally make with the direction of the parent vessel (Fig.3.8.2). It is found [20,21] that in order to minimize the pumping power required to drive the flow through the junction, and again based on the assumtion of Poiseuille flow, optimum branching angles are given by

$$\cos\theta_1 = \frac{(1+\alpha^3)^{4/3} + 1 - \alpha^4}{2(1+\alpha^3)^{2/3}} \tag{3.8.2}$$

$$\cos\theta_2 = \frac{(1+\alpha^3)^{4/3} + \alpha^4 - 1}{2\alpha^2(1+\alpha^3)^{2/3}} \tag{3.8.3}$$

These results indicate that the branch with the smaller diameter makes a larger branching angle (Fig.3.8.2), which is fairly well supported by observations from the cardiovascular system [7–12]. In the limit of a very small branch with $\alpha \approx 0$, the results indicate that the branching angle of the larger branch is near zero while that of the smaller branch is near 90°.

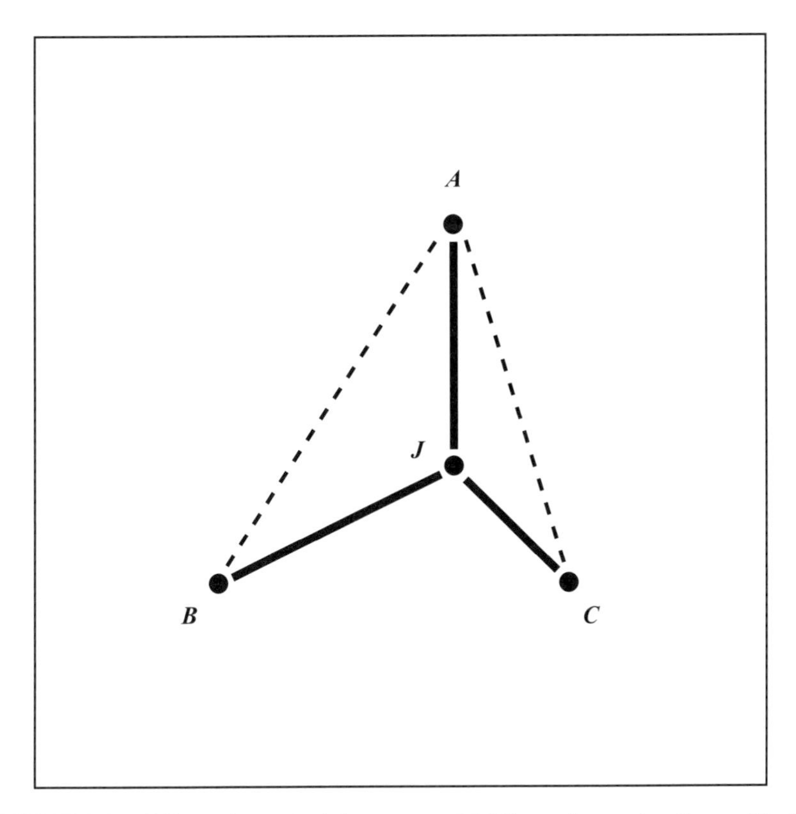

FIGURE 3.8.1. Bifurcation principle: an arterial bifurcation makes it possible for flow from point A to reach points B and C without running two separate tubes from A to B and C. Instead, flow is "shared" from A to some junction point J before dividing into two separate tubes.

This again is supported by observations from the cardiovascular system and is in fact the basis of the common term "small side branch." Other optimality principles for branching angles have been used and have produced qualitatively similar results [22–25].

Arterial trees are made up of a large number of repeated bifurcations to produce a large number of end vessels at the delivering end of the tree. If the tree structure begins with a single tube segment that bifurcates again and again, each time doubling the number of segments involved, the number of such "generations" required to reach a large number of end vessels is fairly small. Only 30 generations produce over 10^9 end segments.

The total cross sectional area available to the flow generally increases at each generation, the increase being mediated by values of the area ratio β greater than 1.0 at individual bifurcations. An accepted estimate in the systemic arterial tree is that the increase in cross–sectional area from the aorta to the capillaries is by a factor of about 1000 [26]. If this increase is assumed to occur over 30 generations in a uniform tree structure in which

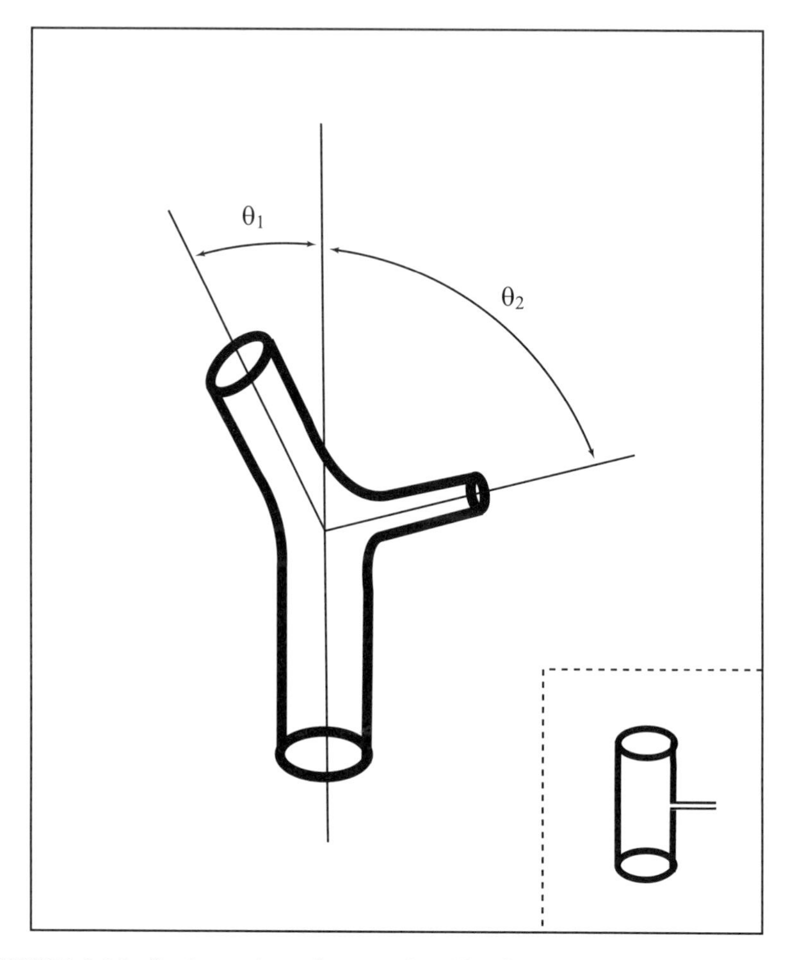

FIGURE 3.8.2. Optimum branching angles. The fluid dynamic efficiency of an arterial bifurcation is affected by the angles at which the two daughter vessels "branch off." Optimally the larger branch makes a smaller branching angle than the smaller branch. In the limit, a small "side branch" comes off at almost 90°, while the branching angle of the larger branch is close to zero (inset).

the value of β is the same at every bifurcation, then an estimate of that value is given by

$$\beta^{30} = 1,000 \tag{3.8.4}$$

which yields

$$\beta = 10^{1/10} \approx 1.2589 \tag{3.8.5}$$

Comparing this value with that obtained from the cube law (Eq.3.7.8), the closeness of the two values is quite remarkable because of the widely different considerations on which they are based.

3.9 Entry Length

Poiseuille flow is an idealized flow based on the assumption of a "very long tube." The tube is assumed to be sufficiently long so that the flow under discussion can be considered to be too far from the tube's entrance to be affected by it. This assumption provides the basis for a major simplification of the governing equations and hence the basis of the solution of these simplified equations that yield Poiseuille flow.

In reality, when flow enters a tube, it takes some distance downstream before it reaches the idealized state of fully developed Poiseuille flow. That distance is referred to as the "entry length" (Fig.3.1.1). An inherent difficulty with the computation or measurement of this length is that the progression toward the idealized state of fully developed Poiseuille flow is an *asymptotic* process. Strictly, the flow never reaches that state.

For practical purposes, however, flow in a tube becomes *very nearly fully developed* at a finite distance from the tube's entrance. That distance has been determined both theoretically and experimentally and is used as the entry length for practical purposes, although in both cases the determination depends critically on the definition of "very nearly fully developed" [27–29]. A criterion that has been commonly used is that the flow is deemed fully developed at such distance from the entrance where the centerline velocity has reached 99% of its Poiseuille flow value. Entry length results are usually based on this criterion, although other criteria based on integral properties of the velocity profile have also been used. The value of the entry length also depends on the form in which flow enters the tube. The assumption usually made here is that the flow enters uniformly.

With these considerations in mind the generally accepted value of the entry length l_e in a tube of diameter d in which the flow at entrance is uniform with velocity U, is given by [27–29]

$$\frac{l_e}{d} = 0.04 R_d \tag{3.9.1}$$

where R_d is the Reynolds number based on the diameter of the tube, namely,

$$R_d = \frac{\rho U d}{\mu} \tag{3.9.2}$$

where ρ, μ are density and viscosity of the fluid, respectively. Thus at $R_d = 1000$ the entry length is equal to 40 tube diameters.

In the cardiovascular system flow is rarely in a very long tube. The vascular tree is made up of vascular segments, which on average have a length to diameter ratio of about 10 and range anywhere from a minimum near zero to a maximum of 35–40 (Fig.3.9.1). Flow in each segment typically does not have the entry length required for that segment. However, the flow is entering and reentering these segments not as uniform flow but as *partially developed flow.*

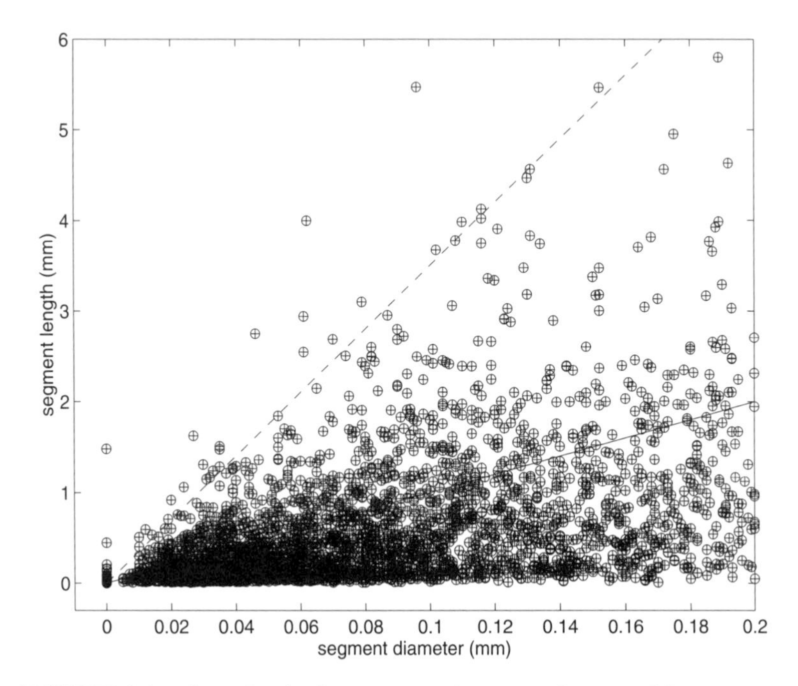

FIGURE 3.9.1. Length of tube segments in a vascular tree. Measurements, here taken from the vascular tree of the human heart, indicate that while there is no apparent correlation between the length and diameter of vessel segments, the *ratio* of length to diameter seems to have a maximum value of about 35 (dashed line) and an average of about 10 (solid line). From [36].

When flow entering a tube is partially developed, or in any case not uniform, its entry length is usually shorter than the corresponding entry length for uniform entry flow [30]. Thus while flow in the arterial tree may not have sufficient run to become fully developed in each tube segment, *it may become increasingly developed as it hops from one segment to the next.* Furthermore, as flow progresses from the central segments of the arterial tree toward the periphery, both the diameter of vessel segments and the average velocity within them rapidly decrease. The Reynolds number, from a high of about 1000 in the aorta (Eq.1.9.3), decreases rapidly in subsequent segments. The flow in these segments therefore requires smaller and smaller length to diameter ratios to become fully developed. At $R_d = 100$, only four tube diameters are required for flow to become fully developed.

These considerations suggest that while flow may not be fully developed everywhere in the vascular tree, it may be very close to that state in many parts of the tree. The assumption of a fully developed state when considering steady flow in the arterial tree, whether in a local region or in the tree as a whole, may be somewhat justified on this basis. From a practical viewpoint the assumption is in any case a necessary approximation to make the analysis tractable. Analysis based on individual flow development in each of many millions of tube segments is clearly impractical.

If fluid enters a tube uniformly with velocity U, then U must equal the average velocity $\overline{u}_s$ of the Poiseuille flow profile, which it ultimately reaches. Velocity along the centerline of the tube, however, must change from U to the *maximum* velocity in the Poiseuille profile, namely, $u_s(0)$, which is twice the average velocity as determined in (Eq.3.4.4). Thus in the entry region of flow in a tube, fluid near the centerline of the tube must be accelerated, and this acceleration requires extra pumping power to maintain. Analysis based on fully developed flow is therefore likely to underestimate the power required to drive the flow in the vascular tree.

In pulsatile flow the entry flow problem is further complicated by the combination of flow development in space and time [31,32]. When the frequency of pulsation is low, the flow reaches the Poiseuille flow profile at the peak of each cycle *in the fully developed region*, thus the results of fully developed *steady* flow bear some relevance to pulsatile flow in that region. In the entry region the situation is more complicated. In its simplest form the flow at a given distance from the tube's entrance attempts to reach the velocity profile that prevails in steady flow at that location, but the problem is not actually as simple because equations governing the flow in the entrance region are nonlinear.

3.10 Noncircular Cross Section

Fully developed Poiseuille flow in a tube is a singularly efficient form of fluid flow because it combines two unique properties of the tube: its cylindrical form and its circular cross section. Any deviation from either of these two ideal conditions causes the flow to be less efficient. If the tube is bent or kinked, or if its lumen is obstructed in any way, the flow will be less efficient because the space available to the fluid is no longer cylindrical. If the tube is straight and cylindrical but its cross section is noncircular, the flow will be less efficient. Remarkably, *any* noncircular cross section produces less efficient flow [33,34].

While these remarks apply to steady flow, they extend fully to pulsatile flow, hence, the relevance of the discussion to the main subject of this book. Under pulsatile conditions the flow aims to reach the fully developed steady flow profile at the peak of each cycle and, as we shall see later, the extent to which it achieves that aim depends on the pulsation frequency. If the fully developed steady-state flow is not Poiseuille flow because of any deviation from the ideal conditions of a cylindrical tube with circular cross section, the efficiency of pulsatile flow will be reduced in the same way and for the same reasons for which it is reduced in steady flow, and this reduction will be compounded by the effect of frequency [35].

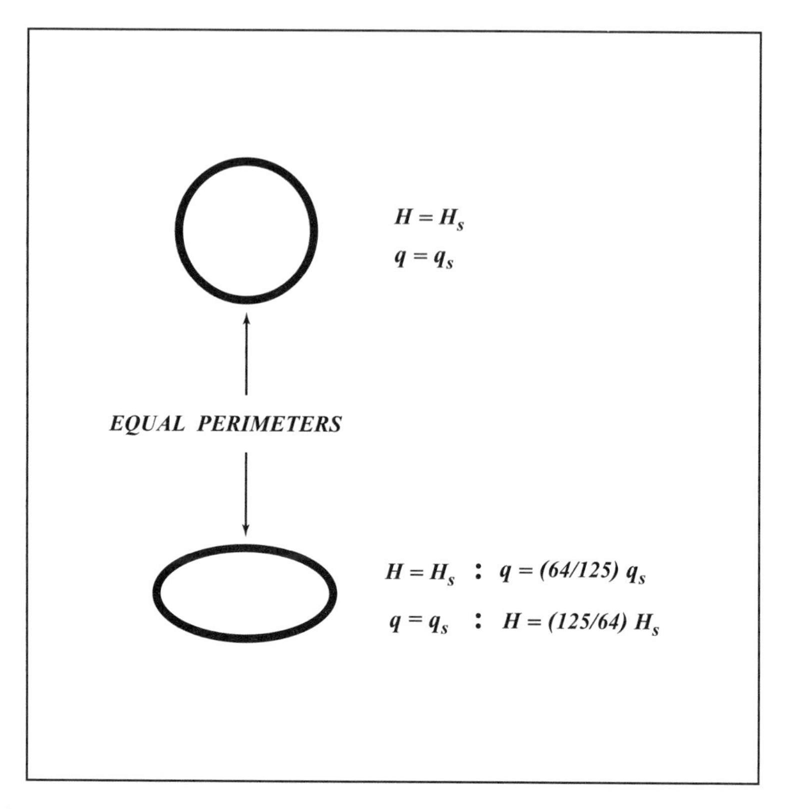

FIGURE 3.10.1. Flow in a tube of elliptic cross section compared with that in a tube of circular cross section of equal perimeter. For the same flow rate in both tubes, the pumping power required to drive the flow in the elliptic tube is higher by a factor of 125/64. For the same pumping power in both tubes, the flow rate in the elliptic tube is lower by a factor of 64/125. The ratio of major to minor axis of the ellipse is 2.0.

In this section we consider briefly steady flow in a tube of elliptic cross section as an example of deviation from the singular case of a tube of circular cross section. An elliptic cross section is particularly relevant in blood flow because it may represent approximately the cross section of a compressed blood vessel.

One important feature of flow in a tube carries over from the case of circular cross section to that of noncircular cross section. If the noncircular tube is straight and cylindrical, the flow remains one directional and the transverse velocity components are identically zero as they are in the case of circular cross section. Under these conditions the governing equations for the flow in a tube of any cross section is the same as that for a tube of circular cross section, which in rectangular Cartesian coordinates x, y, z, with x along the axis of the tube and y, z in the plane of a cross section, takes the form

$$\mu \left(\frac{\partial^2 u_{se}}{\partial y^2} + \frac{\partial^2 u_{se}}{\partial z^2} \right) = k_s \tag{3.10.1}$$

where u_{se}, k_s, μ are as defined previously for the case of a tube of circular cross section (Eq.3.3.4), with subscript 'e' being added here as a reference to the case of elliptic cross section to be considered. To facilitate comparison with the case of circular cross section, the constant pressure gradient k_s is here taken to be the same as that in a tube of circular cross section.

The difference between a circular and a noncircular cross section appears only in the *boundary conditions*. In the present case the no-slip boundary condition must be satisfied on the wall of an *elliptic* tube. If the cross section of the tube is an ellipse of semi-minor and semi-major axes b, c, respectively, the condition is

$$u_{se} = 0 \quad \text{on} \quad \frac{y^2}{b^2} + \frac{z^2}{c^2} = 1 \tag{3.10.2}$$

With this boundary condition plus the condition of symmetry of the flow field, Eq.3.10.1 has the solution

$$u_{se}(y,z) = \frac{k_s}{2\mu} \frac{b^2 c^2}{b^2 + c^2} \left(\frac{y^2}{b^2} + \frac{z^2}{c^2} - 1 \right) \tag{3.10.3}$$

The flow rate and pumping power corresponding to this solution are respectively given by

$$q_{se} = 4 \int_0^c \int_0^{b\sqrt{1-z^2/c^2}} u dy dz = \frac{-k_s \pi}{8\mu} \delta^4 \tag{3.10.4}$$

$$H_{se} = k_s l q_{se} = \frac{8\mu l}{\pi} \frac{q^2}{\delta^4} \tag{3.10.5}$$

where

$$\delta = \left(\frac{2b^3 c^3}{b^2 + c^2} \right)^{1/4} \tag{3.10.6}$$

When $b = c = a$, $\delta = a$, the ellipse becomes a circle of radius a, and the expressions for flow rate and pumping power reduce to those for a tube of circular cross section of radius a.

Flow inefficiency in a tube of noncircular cross section manifests itself in terms of lower flow rate for a given pumping power, or higher pumping power for a given flow rate. For a meaningful comparison between the two cases, either the *areas* or the *perimeters* of the circular and the noncircular cross sections must be made equal. In the case of elliptic cross section this establishes a relation between the radius of the circular cross section and the axes of the elliptic cross section. Comparison of two cross sections of equal perimeter is particularly relevant to the case of compressed blood vessels, and we use it here for illustration.

The perimeter of an ellipse of semi-minor and semi-major axes b, c is equal to that of a circle of radius a if

$$a^2 \approx \frac{(b^2 + c^2)}{2}. \tag{3.10.7}$$

For an ellipse of aspect ratio $c/b = 2$ we have from Eq.3.10.6

$$\delta = \left(\frac{16}{5}\right)^{1/4} b \tag{3.10.8}$$

and if the ellipse is to be compared with a circle of equal perimeter, this gives, from Eq.3.10.7,

$$\delta^4 = \frac{64}{125} a^4 \tag{3.10.9}$$

Thus for a given pumping power the flow rate (Eq.3.10.4) is reduced by a factor of 64/125, while for a given flow rate the pumping power (Eq.3.10.5) is increased by a factor of 125/64 (Fig.3.10.1).

3.11 Problems

1. Many results for flow in a tube are based on "fully developed flow". Explain the defining properties of this flow.

2. List the assumptions on which the equation governing steady or pulsatile flow in a rigid tube (Eq.3.2.9) is based.

3. In the solution for Poiseuille flow, the velocity $\hat{u}_s$ along the axis of the tube is related to the pressure gradient k_s by (Eq.3.4.1)

$$\hat{u}_s = \frac{-k_s a^2}{4\mu}$$

where a is the radius of the tube and μ is viscosity of the fluid. Explain the minus sign in this relation.

4. Show that Eq.3.3.2 for Poiseuille flow in a tube can be derived directly by considering a balance of forces acting on a volume of fluid within the tube.

5. Show that if Eq.3.3.2 is solved for flow in a tube of radius a with *slip-velocity* u^* at the tube wall instead of no-slip boundary condition there, the flow rate would be given by

$$q_s = \pi a^2 u^* - \frac{k_s \pi a^4}{8\mu}$$

Comparison with Eq.3.4.3 shows that slip *increases* the flow rate, as would be expected on physical grounds.

6. The possibility of "slip" at the vessel wall was used in the past to explain the so-called Fahraeus–Lindqvist effect in blood flow, whereby the coefficient of viscosity of blood is found to drop in vessels of smaller diameter. Use the results of the previous exercise to show that if μ is the actual viscosity of the fluid in the presence of slip-velocity u^* and μ^* is what the viscosity would be in the absence of slip, then

$$\frac{\mu}{\mu^*} = 1 - \frac{8\mu u^*}{k_s a^2}$$

7. If the diameter of a tube is reduced by only 10%, all else being unchanged, find in percent the additional power required to drive the same flow rate through it.
8. Show how the area ratio β at an arterial bifurcation is related to the bifurcation index α if the cube law is replaced by the more general law $q \propto a^n$. In the case of a symmetrical bifurcation, compare the values of β for $n = 2, 4$ with that obtained from the cube law.
9. In an arterial tree in which the branches have successively smaller and smaller diameters, consider the consequences of the cube law ($n = 3$) on the way in which the pumping power and wall shear stress will vary at different levels of the tree structure. Compare the results with those of $n = 2$ and $n = 4$.
10. The key design principle of an arterial tree is to avoid running two tubes in parallel where they can be replaced by one tube carrying the combined flow. Find the additional fractional power required for driving flow in two tubes instead of one in terms of the corresponding area ratio, assuming for simplicity that the two tubes are of equal diameter. Show that the result is consistent with that in Eq.3.8.1.
11. Show that the result for entry length in Eq.3.9.1, if supplemented by the cube law, would suggest that the entry length in an arterial tree diminishes rapidly at higher levels of the tree.
12. The result in Eq.3.10.9 indicates that if a tube of circular cross section is compressed so that its cross section becomes elliptical with an aspect ratio of 2.0, the flow rate through the tube is reduced by approximately one half for the same pumping power, or the pumping power is approximately doubled for the same flow rate. Obtain the corresponding results if the aspect ratio of the elliptic cross section is only 1.1.

3.12 References and Further Reading

1. Rouse H, Ince S, 1957. History of Hydraulics. Dover Publications, New York.
2. Tokaty GA, 1971. A History and Philosophy of Fluidmechanics. Foulis, Henley-on-Thames, Oxfordshire.
3. Murray CD, 1926. The physiological principle of minimum work. I. The vascular system and the cost of blood volume. Proceedings of the National Academy of Sciences 12:207–214.

4. Thompson D'AW, 1942. On Growth and Form. Cambridge University Press, Cambridge.
5. Rodbard S, 1975. Vascular caliber. Cardiology 60:4–49.
6. Hutchins GM, Miner MM, Boitnott JK, 1976. Vessel caliber and branch-angle of human coronary artery branch-points. Circulation Research 38:572–576.
7. Zamir M, Medeiros JA, Cunningham TK, 1979. Arterial bifurcations in the human retina. Journal of General Physiology 74:537–548.
8. Zamir M, Brown N, 1982. Arterial branching in various parts of the cardiovascular system. American Journal of Anatomy 163:295–307.
9. Zamir M, Medeiros JA, 1982. Arterial branching in man and monkey. Journal of General Physiology 79:353–360.
10. Mayrovitz HN, Roy J, 1983. Microvascular blood flow: evidence indicating a cubic dependence on arteriolar diameter. American Journal of Physiology 245: H1031–H1038.
11. Zamir M, Phipps S, Langille BL, Wonnacott TH, 1984. Branching characteristics of coronary arteries in rats. Canadian Journal of Physiology and Pharmacology 62:1453–1459.
12. Zamir M, Chee H, 1986. Branching characteristics of human coronary arteries. Canadian Journal of Physiology and Pharmacology 64:661–668.
13. Kassab GS, Rider CA, Tang NJ, Fung YC, Bloor CM, 1993. Morphometry of pig coronary arterial trees. Americal Journal of Physiology 265:H350–H365.
14. Zamir M, 1996. Tree structure and branching characteristics of the right coronary artery in a right-dominant human heart. Canadian Journal of Cardiology 12:593–599.
15. Sherman TF, 1981. On connecting large vessels to small. The meaning of Murray's Law. Journal of General Physiology 78:431–453.
16. Roy AG, Woldenberg MJ, 1982. A generalization of the optimal models of arterial branching. Bulletin of Mathematical Biology 44:349–360.
17. Woldenberg MJ, Horsfield K, 1983. Finding the optimal length for three branches at a junction. Journal of Theoretical Biology 104:301–318.
18. Zamir M, 1981. Three-dimensional aspects of arterial branching. Journal of Theoretical Biology 90:457–476.
19. Zamir M, Wrigley SM, Langille BL, 1983. Arterial bifurcations in the cardiovascular system of a rat. Journal of General Physiology 81:325–335.
20. Murray CD, 1926. The physiological principle of minimum work applied to the angle of branching of arteries. Journal of General Physiology 9:835–841.
21. Zamir M, 1978. Nonsymmetrical bifurcations in arterial branching. Journal of General Physiology 72:837–845.
22. Kamiya A, Togawa T, 1972. Optimal branching structure of the vascular tree. Bulletin of Mathematical Biophysics 34:431–438.

23. Kamiya A, Togawa T, Yamamoto N, 1974. Theoretical relationship between the optimal models of the vascular tree. Bulletin of Mathematical Biology 36:311–323.

24. Uylings HBM, 1977. Optimization of diameters and bifurcation angles in lung and vascular tree structures. Bulletin of Mathematical Biology 39:509–520.

25. Kamiya A, Bukhari R, Togawa T, 1984. Adaptive regulation of wall shear stress optimizing vascular tree function. Bulletin of Mathematical Biology 46:127–137.

26. Burton AC, 1965. Physiology and Biophysics of the Circulation. Year Book Medical Publishers, Chicago.

27. Lew HS, Fung YC, 1970. Entry length into blood vessels at arbitrary Reynolds number. Journal of Biomechanics 3:23–38.

28. Fung YC, 1984. Biodynamics: Circulation. Springer-Verlag, New York.

29. Schlichting H, 1979. Boundary Layer Theory. McGraw-Hill, New York.

30. Camiletti SE, Zamir M, 1984. Entry length and pressure drop for developing Poiseuille flows. Aeronautical Journal 88:265–269.

31. Caro CG, Pedley TJ, Schroter RC, Seed WA, 1978. The Mechanics of the Circulation. Oxford University Press, Oxford.

32. Chang CC, Atabek HB, 1961. The inlet length for oscillatory flow and its effects on the determination of the rate of flow in arteries. Physics in Medicine and Biology 6:303–317.

33. Begum R, Zamir M, 1990. Flow in tubes of non-circular cross sections. In: Rahman M (ed), Ocean Waves Mechanics, Computational Fluid Dynamics and Mathematical Modelling. Computational Mechanics Publications, Southampton.

34. Quadir R, M. Zamir, 1997. Entry length and flow development in tubes of rectangular and elliptic cross sections. In: Rahman M (ed), Laminar and Turbulent Boundary Layers. Computational Mechanics Publications, Southampton.

35. Haslam M, Zamir M, 1998. Pulsatile flow in tubes of elliptic cross sections. Annals of Biomechanics 26:1–8.

36. Zamir M, 1999. On fractal properties of arterial trees. Journal of Theoretical Biology 197:517–526.

4

Pulsatile Flow in a Rigid Tube

4.1 Introduction

Flow in a tube in which the driving pressure varies in time is governed by Eq.3.2.9, namely,

$$\rho\frac{\partial u}{\partial t}+\frac{1}{\rho}\frac{\partial p}{\partial x}=\mu\left(\frac{\partial^2 u}{\partial r^2}+\frac{1}{r}\frac{\partial u}{\partial r}\right)$$

Providing that all the simplifying assumptions on which the equation is based are still valid, the equation provides a forum for a solution in which the pressure p is a function of x and t while the velocity u is a function of r and t. Before obtaining this solution, it is important to reiterate the assumptions on which the equation is based, because these assumptions define the idealized features of the flow that the solution represents.

For the purpose of discussion we consider a specific case in which the driving pressure is oscillatory in time, varying as a trigonometric sine or cosine function. As the pressure rises to its peak, the flow increases gradually in response, and as the pressure falls, the flow follows again. If the change in pressure is very slow, the corresponding change in flow will be almost in phase with it, but if the change in pressure is rapid, the flow will lag behind because of the inertia of the fluid. Because of this lag, the peak that the flow reaches in each cycle will be somewhat short of what it would be in steady Poiseuille flow under a constant driving pressure equal to the peak of the oscillatory pressure.

This loss in peak flow is higher at higher frequency of oscillation of the driving pressure, to the point that at very high frequency the fluid barely moves at all. Only at the other extreme, at very *low* frequency, will the flow rise and fall in phase with the pressure and reach a peak commensurate with the peak pressure at each cycle. In fact at very low frequency the pressure and flow are *instantaneously* what they would be in steady Poiseuille flow. That is, the Poiseuille relation between pressure and flow (Eq.3.4.3) is satisfied at every instant in the oscillatory cycle as they both change during the cycle. Not only the flow rate but the velocity profile at each instant will also be the same as it would be in steady Poiseuille flow under a constant driving pressure equal to the value of the oscillatory pressure at that instant in the oscillatory cycle, at *low* frequency. At *high* frequency the velocity profile fails to reach the full form that it would have reached in Poiseuille flow under the same driving pressure.

The assumptions under which this oscillatory flow takes place are essentially the same as those in steady Poiseuille flow. The cross section of the tube must be circular and axial symmetry must prevail to the effect that velocity and derivatives in the θ direction are zero. Also, the tube must be sufficiently long for the flow field to be fully developed and independent of x, and we saw in Section 3.2 that this requires that the tube be *rigid*. The consequences of these restrictions are far more significant in pulsatile flow than in steady Poiseuille flow.

In order to satisfy these restrictions in pulsatile flow, fluid at different axial positions along the tube must respond *in unison* to the changing pressure, to the effect that the velocity profile is *instantaneously* the same at all axial positions along the tube. As the pressure changes, the velocity profile changes in response, simultaneously at all axial positions along the tube. It is as if the fluid is moving in bulk.

While this feature of the flow is fairly artificial and somewhat "unphysical," it provides an important foundation for the understanding of more realistic forms of pulsatile flow. In fact the classical solution that we present in this chapter and that has provided the basic understanding of pulsatile flow is based on this model of the flow.

To make the model more realistic, the tube must be allowed to be nonrigid. As the pressure changes in a nonrigid tube, the change acts only locally at first because it is able to stretch the tube at that location (Fig.5.1.1). Later the stretched section of the tube recoils and pushes the change in pressure further down the tube. This produces a *wave*, which travels down the tube [1]. *In the case of a rigid tube there is no wave motion.* The flow in a rigid tube rises and falls simultaneously at all axial positions along the tube.

In the presence of wave motion the axial velocity u is a function of not only r and t but also of x, and the radial velocity v is no longer zero, thus Eq.3.2.9 ceases to be valid. More important, the presence of wave motion entails the possibility of *wave reflections*, which introduce further complications in the analysis of the flow. These complications are considered in subsequent chapters. In this chapter we present the classical solution of

Eq.3.2.9, which is based on the idealized model of pulsatile flow in a rigid tube.

4.2 Oscillatory Flow Equations

The pumping action of the heart produces a pressure difference across the arterial tree, which changes rhythmically with that action. It is a characteristic of this driving force that it consists of a constant part that does not vary in time and that produces a steady flow forward as in Poiseuille flow, plus an oscillatory part that moves the fluid only back and forth and that produces zero net flow over each cycle. We shall use the terms "steady" and "oscillatory" to refer to these two components of the flow, respectively, and the term "pulsatile" to refer to the combination of the two.

An important feature of Eq.3.2.9 is that it is *linear* in both the pressure $p(x,t)$ and velocity $u(r,t)$. As a result of this feature the equation can deal with the steady and oscillatory parts of the flow entirely separately and independently of each other. This is a useful breakdown of the problem, because the steady part of the flow has already been dealt with in Chapter 3.

The oscillatory part of the problem can be isolated and dealt with separately, which we do in the present chapter. This part of the problem is mathematically more complicated than the steady-flow part, and in the midst of the analysis to follow it may seem pointless to consider it since it only moves fluid back and forth with no net flow. It is therefore helpful to remember that the reason for which the ocillatory part is dealt with in such great detail is that it represents an important part of the composite pulsatile flow that we are interested in, and it carries with it most of the seminal features of the composite flow.

If the steady and oscillatory parts of the pressure and velocity are identified by subscripts "s" and "ϕ," respectively, then to isolate the oscillatory flow problem we write

$$\begin{aligned} p(x,t) &= p_s(x) + p_\phi(x,t) \\ u(r,t) &= u_s(r) + u_\phi(r,t) \end{aligned} \tag{4.2.1}$$

Substituting these into Eq.3.2.9, we obtain

$$\begin{aligned} &\left\{ \frac{dp_s}{dx} - \mu\left(\frac{d^2u_s}{dr^2} + \frac{1}{r}\frac{du_s}{dr}\right)\right\} \\ + &\left\{ \rho\frac{\partial u_\phi}{\partial t} + \frac{\partial p_\phi}{\partial x} - \mu\left(\frac{\partial^2 u_\phi}{\partial r^2} + \frac{1}{r}\frac{\partial u_\phi}{\partial r}\right)\right\} = 0 \end{aligned} \tag{4.2.2}$$

where terms have been grouped into those that do not depend on time t (first group) and those that do (second group). Because of that difference between them, each group must equal zero separately.

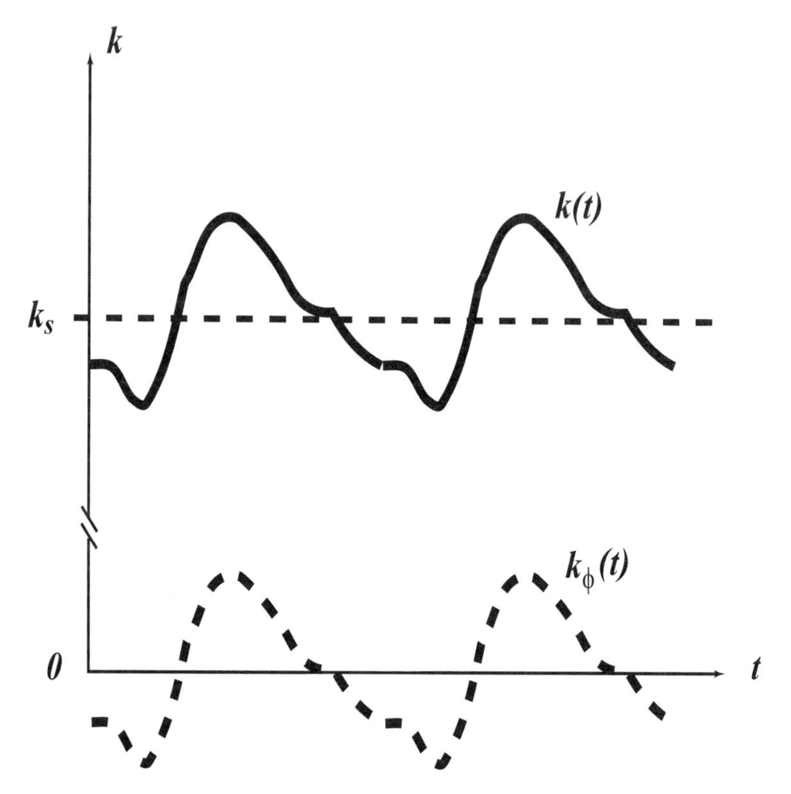

FIGURE 4.2.1. Pulsatile pressure gradient $k(t)$ consists of a constant part k_s and a purely oscillatory part $k_\phi(t)$.

The first equality then leads to Eq.3.3.2 for steady flow, which has already been dealt with, and the second leads to a governing equation for the oscillatory part of the flow, namely,

$$\rho\frac{\partial u_\phi}{\partial t} + \frac{\partial p_\phi}{\partial x} - \mu\left(\frac{\partial^2 u_\phi}{\partial r^2} + \frac{1}{r}\frac{\partial u_\phi}{\partial r}\right) = 0 \qquad (4.2.3)$$

Equations 3.3.2 and 4.2.3 are entirely independent of each other in the sense that the first can be solved for u_s, which has already been done, and the second can be solved separately for u_ϕ, which we do in what follows.

Furthermore, because of the independence of the steady- and oscillatory-flow equations, and for reasons similar to those in steady flow, the relation between pressure *gradients*, from Eq.4.2.1, is

$$k(t) = k_s + k_\phi(t) \qquad (4.2.4)$$

where

$$k(t) = \frac{\partial p}{\partial x}$$

$$k_s = \frac{dp_s}{dx}$$

$$k_\phi(t) = \frac{\partial p_\phi}{\partial x} \tag{4.2.5}$$

Thus $k(t)$ is the "total" pressure gradient in pulsatile flow, k_s is its steady part, and k_ϕ is its purely oscillatory part, as illustrated in Fig.4.2.1

In steady Poiseuille flow the pressure gradient term in the governing equation (Eq.3.3.2) is found to be constant, independent of x, essentially because all other terms in the equation are functions of r only, while the pressure is a function of x only. Similarly, in oscillatory flow the pressure gradient term in the governing equation (Eq.4.2.3) is independent of x, for the same reasons, but here it can be a function of time t.

As k_s in steady Poiseuille flow, k_ϕ here is a measure of the pressure difference $\triangle p_\phi$ between the two ends of the tube, which in the present case is a function of time. In analogy with Eq.3.4.9, here

$$\triangle p_\phi = p_\phi(l,t) - p_\phi(0,t) = k_\phi(t) l \tag{4.2.6}$$

where l is the length of the tube and $p_\phi(x,t)$ is the pressure at axial position x along the tube and at time t. The governing equation for oscillatory flow, from Eq.4.2.3, is then of the form

$$\mu \left(\frac{\partial^2 u_\phi}{\partial r^2} + \frac{1}{r}\frac{\partial u_\phi}{\partial r} \right) - \rho \frac{\partial u_\phi}{\partial t} = k_\phi(t) \tag{4.2.7}$$

4.3 Fourier Analysis

For the oscillatory problem to be physically determined, the way in which the driving pressure varies in time must be given. That is, in order to solve Eq.4.2.7 for $u_\phi(r,t)$, $k_\phi(t)$ must be specified. We are interested in a solution for which $k_\phi(t)$ is an oscillatory function of time, to model the oscillatory pressure produced by the heart discussed in the previous section. However, the oscillatory pressure produced by the heart is not a simple function of time for which a direct solution of Eq.4.2.7 is possible.

One way of dealing with this difficulty is to specify $k_\phi(t)$ *numerically* as a function of time, then Eq.4.2.7 would have to be solved numerically. Another way is offered by the fact that any periodic function can be expressed as a sum of sine and cosine functions, known as a *Fourier series.*

Briefly, a function $f(t)$ is periodic if

$$f(t+T) = f(t) \tag{4.3.1}$$

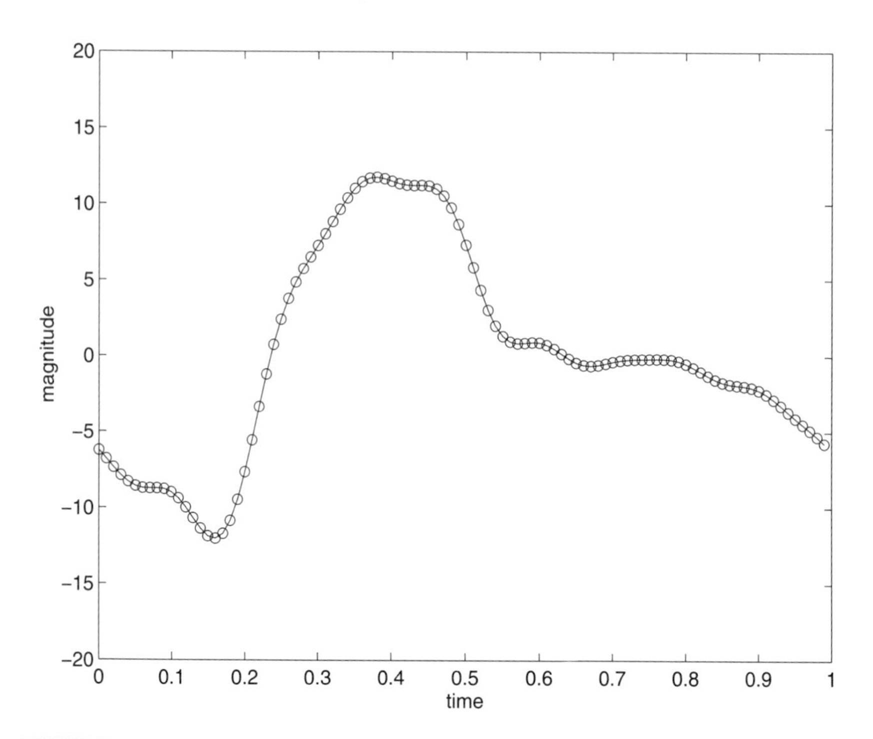

FIGURE 4.3.1. Typical composite pressure waveform produced by the heart. Only the oscillatory part of the wave is shown, the constant part has been removed.

where T is the "period" of the function. A periodic function can be represented by the Fourier series [2]

$$\begin{aligned} f(t) &= \sum_{0}^{\infty} A_n \cos\left(\frac{2n\pi t}{T}\right) + \sum_{1}^{\infty} B_n \sin\left(\frac{2n\pi t}{T}\right) \\ &= A_0 + A_1 \cos\left(\frac{2\pi t}{T}\right) + A_2 \cos\left(\frac{4\pi t}{T}\right) + \ldots \\ &\qquad + B_1 \sin\left(\frac{2\pi t}{T}\right) + B_2 \sin\left(\frac{4\pi t}{T}\right) + \ldots \end{aligned} \tag{4.3.2}$$

where the A's and the B's are constants that are determined by the specific characteristics of $f(t)$ and are given by

$$\begin{aligned} A_0 &= \frac{1}{2\pi} \int_0^{2\pi} f(t)dt \\ A_n &= \frac{1}{\pi} \int_0^{2\pi} f(t) \cos\left(\frac{2n\pi t}{T}\right) dt \end{aligned}$$

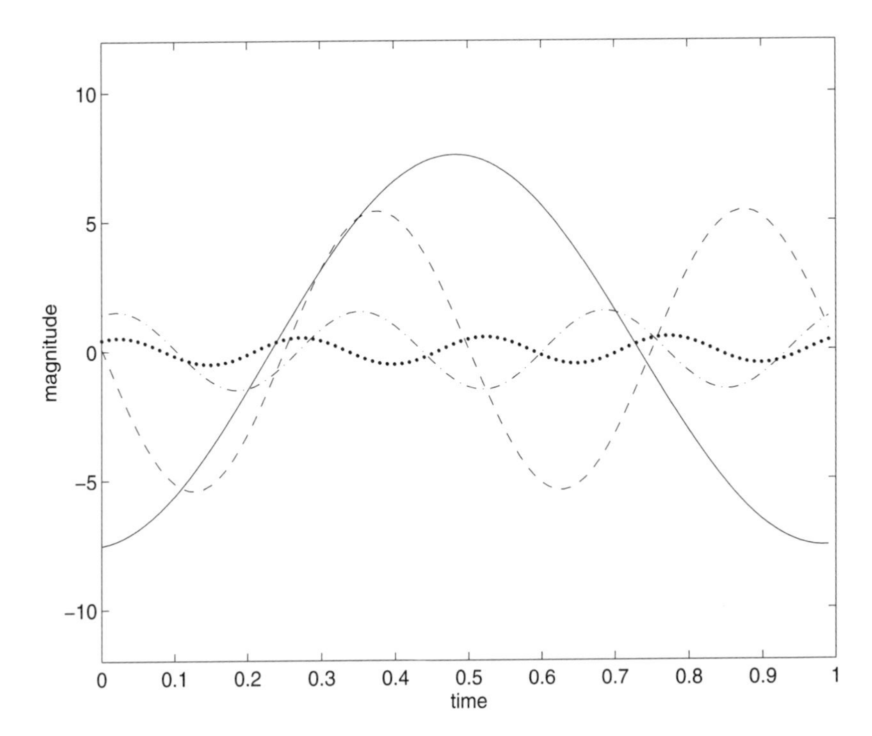

FIGURE 4.3.2. The first four harmonics of the composite waveform shown in Fig.4.3.1. The first harmonic (solid line) has the same frequency as the composite wave, then each subsequent harmonic has double the frequency of its predecessor. Note the diminishing amplitude of higher harmonics.

$$B_n = \frac{1}{\pi} \int_0^{2\pi} f(t) \sin\left(\frac{2n\pi t}{T}\right) dt \tag{4.3.3}$$

Thus any arbitrary form of $k_\phi(t)$ in Eq.4.2.7, so long as it is a *periodic* function, can be expressed as a Fourier series consisting of a sum of sines and cosines known as its "harmonics" [2]. It will be noted from Eq.4.3.2 that the nature of these harmonics is such that the first harmonic has the same frequency as the original waveform, the second has double that frequency, and so on. Thus the original composite waveform can be decomposed into a "recipe" of sine and cosine waves of increasing frequency and, as we shall see, diminishing amplitude. This makes the method of great practical use because only a relatively small number of harmonics (10 or less) are usually required to represent the composite wave with reasonable accuracy. An example of an oscillatory waveform produced by the heart is shown in Fig.4.3.1, with its first four harmonics shown in Fig.4.3.2.

The advantage of a solution based on this approach is twofold. First, since Eq.4.2.7 is *linear* in u_ϕ, a sum of sine and cosine terms on the right can be taken only one term at a time, then adding the results. Thus the equation, in effect, need be solved with only one sine or cosine term on the right. Second, Eq.4.2.7 has an exact analytical solution when $k_\phi(t)$ is a sine or cosine function, which has many advantages over a numerical solution.

Added to these advantages is the fact that in most practical cases the infinite Fourier series in Eq.4.3.2 can be approximated by a sum of only 10 or so terms. And the process of decomposition of the solution and re-synthesis of the results have been automated by many computer programs, which handle all the tedious details [3]. Thus the classical solution that we present in what follows, which is highly idealized in that it is based on a driving pressure which varies as a sine or cosine function, is in fact an essential element of a more realistic solution in which the pressure varies as a more general oscillatory function.

4.4 Complex Pressure Gradient and Bessel Equation

Solution of Eq.4.2.7 with the oscillatory pressure gradient being taken as a sine or cosine function is much simplified analytically if, instead of using one or the other, their complex combination in exponential form is used, that is, taking

$$k_\phi(t) = k_s e^{i\omega t} = k_s(\cos\omega t + i\sin\omega t) \tag{4.4.1}$$

where $i = \sqrt{-1}$. Because Eq.4.2.7 is linear, a solution with this choice of k_ϕ will actually consist of the sum of *two* solutions, one for which $k_\phi(t) = k_s\cos\omega t$ and another for which $k_\phi(t) = k_s\sin\omega t$. The first is obtained by taking the real part of the solution and the second by taking the imaginary part. The combined solution is complex because of this choice of $k_\phi(t)$.

For the purpose of comparison with steady Poiseuille flow in which the driving pressure gradient is constant, k_s, the amplitude of the oscillatory pressure gradient in Eq.4.4.1 is taken as k_s. In this way the peak value of the oscillatory flow rate and peak form of the oscillatory velocity profile can be compared with the flow rate and velocity profile in steady Poiseuille flow with constant pressure gradient k_s.

The governing equation for oscillatory flow with this choice of pressure gradient is then, using Eqs.4.4.1 and 4.2.7,

$$\frac{\partial^2 u_\phi}{\partial r^2} + \frac{1}{r}\frac{\partial u_\phi}{\partial r} - \frac{\rho}{\mu}\frac{\partial u_\phi}{\partial t} = \frac{k_s}{\mu}e^{i\omega t} \tag{4.4.2}$$

The form of the equation admits a solution by separation of variables, that is, by a decomposition of $u_\phi(r,t)$ into one part that depends on r only and one on t only. Furthermore, the form of the equation and the exponential form of the function of time on the right-hand side together dictate that the part of u_ϕ which depends on t must have the same exponential form as that on the right-hand side. The separation of variables thus arrived at is

$$u_\phi(r,t) = U_\phi(r)e^{i\omega t} \tag{4.4.3}$$

Upon substitution in Eq.4.4.2, the factor $e^{i\omega t}$ cancels throughout, leaving an ordinary differential equation for $U_\phi(r)$ only, namely,

$$\frac{d^2U_\phi}{dr^2} + \frac{1}{r}\frac{dU_\phi}{dr} - \frac{i\Omega^2}{a^2}U_\phi = \frac{k_s}{\mu} \tag{4.4.4}$$

where a is the tube radius and Ω is an important nondimensional parameter, given by

$$\Omega = \sqrt{\frac{\rho\omega}{\mu}}a \tag{4.4.5}$$

We shall see later that the value of Ω has a significant effect on the form of the solution.

It is clear that Eq.4.4.4, being an *ordinary* differential equation, is considerably simpler than Eq.4.4.2. In fact Eq.4.4.4 is a form of Bessel equation, which has a standard solution as we see in the next section [4,5].

It is important to note that this simplification of the problem has been achieved primarily by the simple choice of $k_\phi(t)$ in Eq.4.4.1. This simplified form of the problem is fundamental, however, and is in fact a prerequisite for dealing with more complicated forms of $k_\phi(t)$ as discussed in the previous section. It is also important to note that while the pressure gradient $k_\phi(t)$ in Eq.4.4.1 and the velocity $u_\phi(r,t)$ in Eq.4.4.3 appear to have the same oscillatory form in the time variable t, this does not mean that the pressure gradient and velocity are actually in phase with each other. The reason for this is that the other part of the velocity in Eq.4.4.3, namely, $U_\phi(r)$, is a complex entity as can be anticipated from the presence of i in its governing equation (Eq.4.4.4) and as we see in the next section. The product of this complex entity with $e^{i\omega t}$ in Eq.4.4.3 alters the phases of the real and imaginary parts of the velocity $u_\phi(r,t)$ so that in general they are not the same as those of the real and imaginary parts of the pressure gradient $k_\phi(t)$.

4.5 Solution of Bessel Equation

Equation 4.4.4 is a form of Bessel equation that has a known general solution [4,5], namely,

$$U_\phi(r) = \frac{ik_s a^2}{\mu\Omega^2} + AJ_0(\zeta) + BY_0(\zeta) \tag{4.5.1}$$

where A, B are arbitrary constants and J_0, Y_0 are Bessel functions of order zero and of the first and second kind, respectively (Figs.4.5.1,2), satisfying

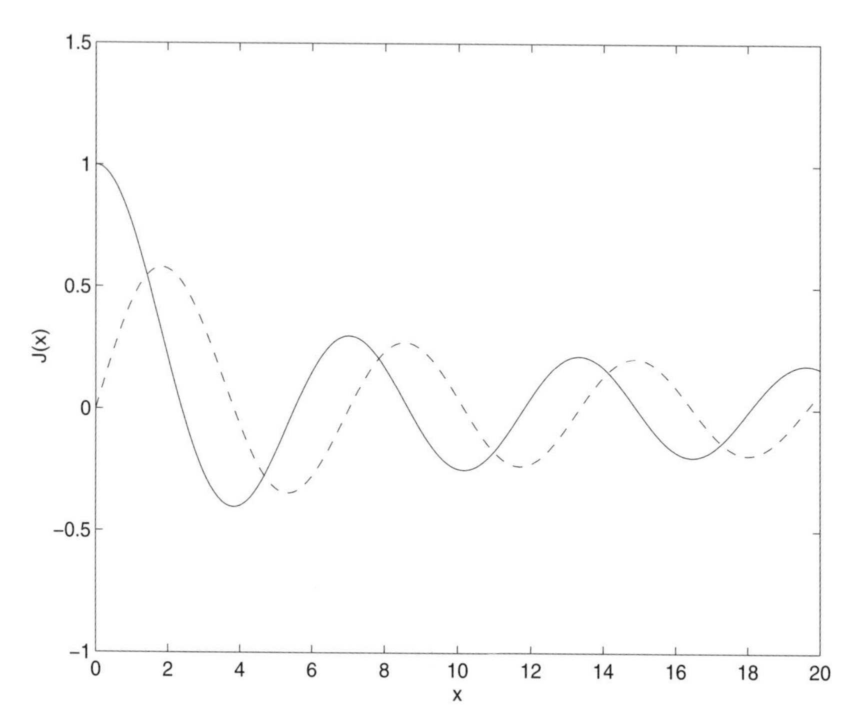

FIGURE 4.5.1. Bessel functions of the first kind, of order zero [$J_0(x)$, solid line] and order one [$J_1(x)$, dashed line], where x is real.

the standard Bessel equation

$$\frac{d^2 J_0}{d\zeta^2} + \frac{1}{\zeta}\frac{dJ_0}{d\zeta} + J_0 = 0 \tag{4.5.2}$$

$$\frac{d^2 Y_0}{d\zeta^2} + \frac{1}{\zeta}\frac{dY_0}{d\zeta} + Y_0 = 0 \tag{4.5.3}$$

The new variable ζ is a complex variable related to the radial coordinate r by

$$\zeta(r) = \Lambda\frac{r}{a} \tag{4.5.4}$$

where Λ is a complex frequency parameter related to the nondimensional frequency Ω by

$$\Lambda = \left(\frac{i-1}{\sqrt{2}}\right)\Omega \tag{4.5.5}$$

Because of this relation, and because Λ and Ω appear explicitly and implicitly in all elements of the solution, the numerical value of the nondimensional frequency Ω has a key effect on the detailed characteristics of the flow field.

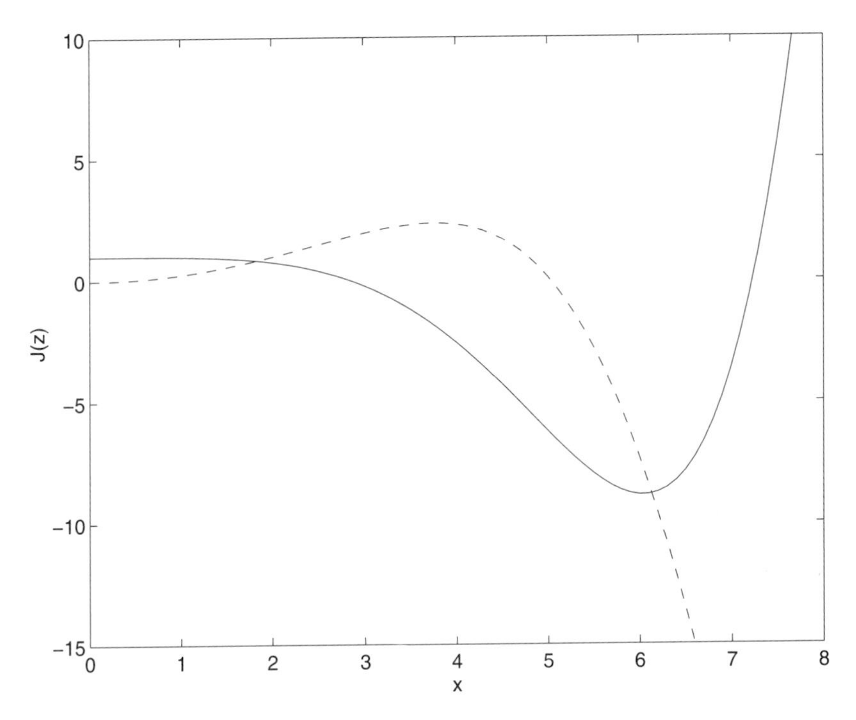

FIGURE 4.5.2. Real part (solid line) and imaginary part (dashed line) of $J_0(z)$, where $z = (i-1)x/2^{1/2}$.

Substituting the solution (Eq.4.5.1) into the governing equation (Eq.4.4.4) readily verifies that the governing equation is satisfied. In fact, it can be readily verified that the first term on the right-hand side of Eq.4.5.1 represents a particular solution of Eq.4.4.4 since it satisfies that equation and it does not contain an arbitrary constant. Then it can be verified that each of the second and third terms on the right-hand side of Eq.4.5.1 satisfies the *homogeneous* form of the governing equation, that is,

$$\frac{d^2U_\phi}{dr^2} + \frac{1}{r}\frac{dU_\phi}{dr} - \frac{i\Omega^2}{a^2}U_\phi = 0 \tag{4.5.6}$$

Substituting $U_\phi = AJ_0$ into this equation produces Eq.4.5.2, which is known to be valid by definition of J_0, and similarly for Y_0. Furthermore, it is known from the properties of J_0 and Y_0 that they are *independent* of each other, that is, one cannot be expressed in terms of the other. Therefore the three terms on the right-hand side of Eq.4.5.1 represent the required elements of the general solution of Eq.4.4.4, namely, two independent solutions of the homogeneous equation (Eq.4.5.6) and a particular solution of the full equation (Eq.4.4.4).

The boundary conditions that the solution must satisfy for flow in a tube are no-slip at the tube wall and finite velocity along the axis of the tube, that is,

$$U_\phi(a) = 0 \tag{4.5.7}$$

$$|U_\phi(0)| < \infty \tag{4.5.8}$$

These provide the required conditions for determining the constants A, B in the solution. It is known from the properties of $Y_0(\zeta)$ that it becomes infinite as $\zeta \to 0$ [4,5], which occurs on the axis of the tube where $r = 0$, thus the boundary condition in Eq.4.5.8 leads to

$$B = 0 \tag{4.5.9}$$

and the first boundary condition then gives

$$A = \frac{-ik_s a^2}{\mu\Omega^2 J_0(\Lambda)} \tag{4.5.10}$$

noting from Eq.4.5.4 that

$$\zeta(a) = \Lambda \tag{4.5.11}$$

With these values of A, B the solution for U_ϕ is finally

$$U_\phi = \frac{ik_s a^2}{\mu\Omega^2}\left(1 - \frac{J_0(\zeta)}{J_0(\Lambda)}\right) \tag{4.5.12}$$

4.6 Oscillatory Velocity Profiles

The solution of Eq.4.4.2 for the oscillatory flow velocity $u_\phi(r,t)$ is now complete. Using Eqs.4.4.3 and 4.5.12, we have

$$u_\phi(r,t) = \frac{ik_s a^2}{\mu\Omega^2}\left(1 - \frac{J_0(\zeta)}{J_0(\Lambda)}\right) e^{i\omega t} \tag{4.6.1}$$

This is a classical solution for oscillatory flow in a rigid tube, obtained in different forms and at different times by Sexl [6], Womersley [7], Uchida [8], and discussed at some length by McDonald [9] and Milnor [10].

The first element of the solution is a constant coefficient whose value depends on the amplitude of the pressure gradient k_s, radius of the tube a, viscosity of the fluid μ, and the frequency of oscillation Ω. The second element, inside the large parentheses, is a function of r that describes the velocity profile in a cross section of the tube. The third element is a function of time, which multiplies and therefore modifies the velocity profile as time changes within the oscillatory cycle, thus producing a sequence of oscillatory velocity profiles.

To compare these oscillatory profiles with the constant parabolic profile in Poiseuille flow, we consider steady and oscillatory flows in tubes of the same radius a, and take k_s in Eq.4.6.1 to be both the constant pressure

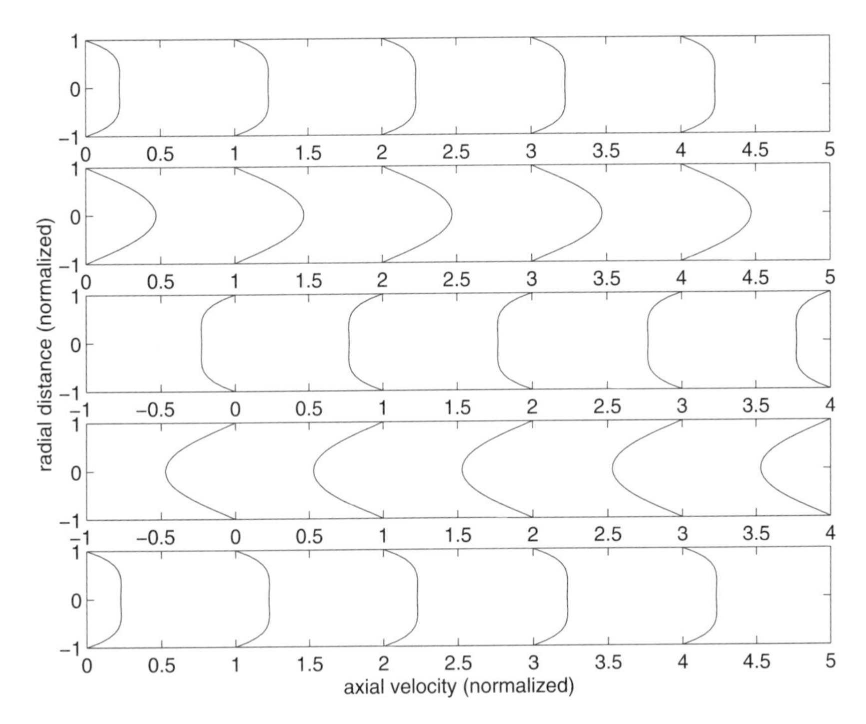

FIGURE 4.6.1. Oscillatory velocity profiles in a rigid tube with frequency parameter $\Omega = 3.0$ and corresponding to the *real* part of the pressure gradient, namely, $k_s \cos \omega t$. The panels represent the profiles at different phase angles (ωt) within the oscillatory cycle, with $\omega t = 0$ in the top panel then increasing by 90° for subsequent panels.

gradient in the Poiseuille flow case and the *amplitude* of the oscillatory pressure gradient in the oscillatory flow case. To further facilitate the comparison, the oscillatory flow velocity $u_\phi(r,t)$ is divided by the maximum velocity in Poiseuille flow $\hat{u}_s$, using Eqs.3.4.1 and 4.5.5, to get

$$\frac{u_\phi(r,t)}{\hat{u}_s} = \frac{-4}{\Lambda^2}\left(1 - \frac{J_0(\zeta)}{J_0(\Lambda)}\right) e^{i\omega t} \tag{4.6.2}$$

This nondimensional form of the oscillatory velocity has the convenient scale in which a value of 1.0 represents a velocity equal to the maximum velocity in the corresponding Poiseuille flow case.

Because of the complex form of the driving pressure gradient (Eq.4.4.1) for which the solution was obtained, Eq.4.6.2 above actually represents two distinct solutions: one for which the pressure gradient varies as the real part of k_ϕ, namely, $\cos \omega t$, and another for which the gradient varies as the imaginary part of k_ϕ, namely, $\sin \omega t$. It is convenient to introduce the following notation for the real and imaginary parts of the velocity and pressure:

$$k_\phi = k_{\phi R} + ik_{\phi I} \tag{4.6.3}$$

$$= k_s(\cos\omega t + i\sin\omega t) \tag{4.6.4}$$

thus

$$k_{\phi R} = k_s\cos\omega t, \quad k_{\phi I} = k_s\sin\omega t \tag{4.6.5}$$

The corresponding velocities are the real and imaginary parts of u_ϕ, that is,

$$u_\phi = u_{\phi R} + iu_{\phi I} = U_\phi e^{i\omega t} \tag{4.6.6}$$

$$= U_\phi(\cos\omega t + i\sin\omega t) \tag{4.6.7}$$

It is important to note that the real and imaginary parts of the velocity *do not* vary as $\cos\omega t$ and $\sin\omega t$ because in the expression for U_ϕ (Eq.4.5.12) the quantities $\Lambda, \zeta, J_0(\zeta), J_0(\Lambda)$ are all complex. Thus the real and imaginary parts of the velocity in these expressions are generally different from those of the pressure gradient, hence a phase difference exists between the oscillatory pressure gradient and the oscillatory velocity profiles that it produces.
If the real and imaginary parts of U_ϕ are denoted by $U_{\phi R}$ and $U_{\phi I}$, respectively, that is,

$$\frac{U_{\phi R}}{\hat{u}_s} = \Re\left\{\frac{-4}{\Lambda^2}\left(1 - \frac{J_0(\zeta)}{J_0(\Lambda)}\right)\right\} \tag{4.6.8}$$

$$\frac{U_{\phi I}}{\hat{u}_s} = \Im\left\{\frac{-4}{\Lambda^2}\left(1 - \frac{J_0(\zeta)}{J_0(\Lambda)}\right)\right\} \tag{4.6.9}$$

and

$$U_\phi = U_{\phi R} + iU_{\phi I} \tag{4.6.10}$$

then Eq.4.6.7 becomes

$$u_\phi = (U_{\phi R} + iU_{\phi I})(\cos\omega t + i\sin\omega t) \tag{4.6.11}$$

and the real and imaginary parts of u_ϕ are given by

$$u_{\phi R} = U_{\phi R}\cos\omega t - U_{\phi I}\sin\omega t \tag{4.6.12}$$

$$u_{\phi I} = U_{\phi I}\cos\omega t + U_{\phi R}\sin\omega t \tag{4.6.13}$$

To facilitate the computation of the real or imaginary parts of the velocity, tables of the complex values of $J_0(\zeta)$ are provided in Appendix A.

It is clear from the expressions in Eq.4.6.1 that the shape of the oscillatory velocity profiles will depend critically on the frequency of oscillation

ω, since ω determines the values of the nondimensional frequency Ω, the complex frequency Λ, the complex variable ζ, and ultimately the Bessel function J_0 and oscillatory flow velocity u_ϕ. A set of oscillatory velocity profiles corresponding to $\Omega = 3$ and to the *real* part of the pressure gradient, namely, $k_s \cos\omega t$, is shown in Fig.4.6.1.

It is observed that velocity profiles oscillate between a peak profile in the forward direction and a peak profile in the backward direction, but neither the phase nor amplitude of these peak profiles correspond with the peaks of the oscillatory pressure. The first because forward and backward peaks of the pressure gradient, being $k_s \cos\omega t$, occur at $\omega t = 0°, 180°$, while the corresponding peak velocity profiles are seen to occur at approximately $\omega t = 90°, 270°$. Thus the oscillatory velocity *lags* the oscillatory pressure, clearly because of the inertia of the fluid. The second because maximum velocity in the peak velocity profile is less than 1.0, which means that it is less than the maximum velocity in the corresponding Poiseuille profile.

4.7 Oscillatory Flow Rate

Volumetric flow rate q_ϕ in oscillatory flow through a tube is obtained by integrating the oscillatory velocity profile over a cross section of the tube. Since the oscillatory velocity $u_\phi(r,t)$ is a function of r and t, the result is a function of time given by

$$q_\phi(t) = \int_0^a 2\pi r u_\phi(r,t)dr \tag{4.7.1}$$

Using the solution for $u_\phi(r,t)$ in Eq.4.6.1, this becomes

$$q_\phi(t) = \frac{2\pi i k_s a^2}{\mu\Omega^2} e^{i\omega t} \int_0^a r\left(1 - \frac{J_0(\zeta)}{J_0(\Lambda)}\right) dr \tag{4.7.2}$$

The integral on the right-hand side is evaluated as follows:

$$\begin{aligned} \int_0^a r\left(1 - \frac{J_0(\zeta)}{J_0(\Lambda)}\right) dr &= \frac{a^2}{\Lambda^2 J_0(\Lambda)} \int_0^\Lambda (J_0(\Lambda) - J_0(\zeta))\zeta d\zeta \\ &= \frac{a^2}{2}\left(1 - \frac{2J_1(\Lambda)}{\Lambda J_0(\Lambda)}\right) \end{aligned} \tag{4.7.3}$$

where J_1 is a Bessel function of the first order and first kind, related to J_0 by

$$\int \zeta J_0(\zeta)d\zeta = \zeta J_1(\zeta) \tag{4.7.4}$$

Thus the oscillatory flow rate is finally given by

$$q_\phi(t) = \frac{i\pi k_s a^4}{\mu\Omega^2}\left(1 - \frac{2J_1(\Lambda)}{\Lambda J_0(\Lambda)}\right) e^{i\omega t} \tag{4.7.5}$$

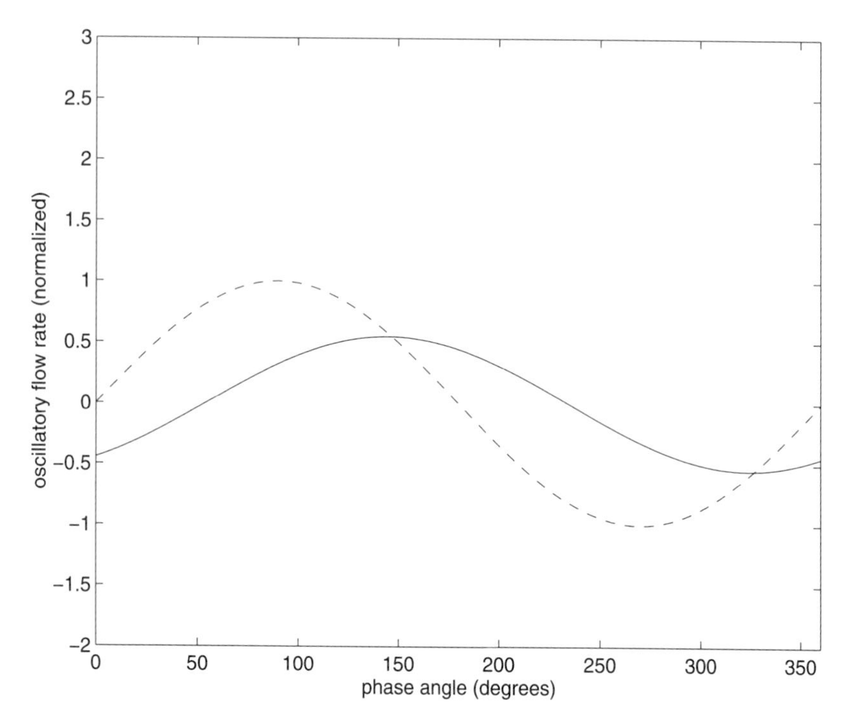

FIGURE 4.7.1. Variation of the oscillatory flow rate q_ϕ within an oscillatory cycle (solid line), compared with variation of the corresponding pressure gradient, in this case $k_{\phi I} = k_s \sin \omega t$ (dashed line). Peak flow occurs later than peak pressure, that is, the flow wave *lags* the pressure wave. Also, normalized peak flow is less than 1.0, which means that peak flow is less than the corresponding Poiseuille flow rate under the same pressure gradient.

The net flow rate Q_ϕ over one oscillatory cycle is given by

$$\begin{aligned} Q_\phi &= \int_0^T q_\phi(t)dt \\ &= \frac{i\pi k_s a^4}{\mu\Omega^2}\left(1 - \frac{2J_1(\Lambda)}{\Lambda J_0(\Lambda)}\right)\int_0^T (\cos\omega t + i\sin\omega t)dt \\ &= \frac{i\pi k_s a^4}{\mu\Omega^2}\left(1 - \frac{2J_1(\Lambda)}{\Lambda J_0(\Lambda)}\right)\int_0^{2\pi} (\cos\theta + i\sin\theta)d\theta \\ &= 0 \end{aligned} \tag{4.7.6}$$

where $T = 2\pi/\omega$ is the period of oscillation. The result confirms that in oscillatory flow the fluid moves only back and forth, with no net flow in either direction.

To examine the variation of flow rate within the oscillatory cycle, it is convenient to put Eq.4.7.5 into nondimensional form, as was done for the velocity in the previous section. Dividing Eq.4.7.5 through by the corresponding flow rate in steady Poiseuille flow, namely, q_s in Eq.3.4.3, gives

$$\frac{q_\phi(t)}{q_s} = \frac{-8}{\Lambda^2}\left(1 - \frac{2J_1(\Lambda)}{\Lambda J_0(\Lambda)}\right) e^{i\omega t} \qquad (4.7.7)$$

The ratio represents the oscillatory flow rate scaled in terms of the corresponding flow rate in Poiseuille flow, thus a value of 1.0 represents a flow rate equal to that in Poiseuille flow in the same tube and under a constant pressure gradient equal to k_s. Numerical computations require values of $J_1(\Lambda)$ in addition to those of $J_0(\Lambda)$. A table of these is given in Appendix A. It is clear from the expression on the right-hand side of Eq.4.7.7 that the oscillatory flow rate depends heavily on the frequency of oscillation. It should also be noted that the expression is *complex*, its real part corresponding to the real part of the pressure gradient, and its imaginary part corresponding to the imaginary part of the pressure gradient.

Variation of $q_\phi(t)/q_s$ over one oscillatory cycle, at a moderate frequency, $\Omega = 3.0$, is shown in Fig.4.7.1. It is seen that the flow rate oscillates between a peak in the forward direction and a peak in the backward direction. At this frequency the value of the peak is decidedly less than 1.0, which means that the flow rate falls short of reaching the corresponding flow rate in Poiseuille flow under a constant pressure gradient equal to k_s, which is the peak value of the oscillatory pressure gradient. The reason for this is clearly the inertia of the fluid, which must be accelerated to peak flow in each cycle. We shall see later that this effect intensifies as the frequency of oscillation increases, the fluid having greater and greater difficulty reaching a peak flow commensurate with that in Poiseuille flow.

4.8 Oscillatory Shear Stress

In oscillatory flow, as fluid moves back and forth in response to the oscillatory pressure gradient, the shear stress exerted by the fluid on the tube wall varies accordingly as a function of time given by

$$\tau_\phi(t) = -\mu\left(\frac{\partial u_\phi(r,t)}{\partial r}\right)_{r=a} \qquad (4.8.1)$$

It is important to note that, since this shear is produced by only the oscillatory velocity u_ϕ, it is *in addition* to that produced by the steady velocity u_s when present.

Using the solution for $u_\phi(r,t)$ in Eq.4.6.1, this becomes

$$\tau_\phi(t) = -\frac{ik_s a^2}{\Omega^2}\left\{\frac{d}{dr}\left(1 - \frac{J_0(\zeta)}{J_0(\Lambda)}\right)\right\}_{r=a} e^{i\omega t}$$

$$= -\frac{ik_s a^2}{\Omega^2}\left\{\frac{d}{d\zeta}\left(1-\frac{J_0(\zeta)}{J_0(\Lambda)}\right)\right\}_{\zeta=\Lambda}\frac{\Lambda}{a}e^{i\omega t}$$

$$= -\frac{k_s a}{\Lambda}\left(\frac{J_1(\Lambda)}{J_0(\Lambda)}\right)e^{i\omega t} \tag{4.8.2}$$

where the relations between ζ,Λ and Ω in Eqs.4.5.4,5 have been used, as well as the following relation between Bessel functions of the first and zeroth order [4,5]:

$$\frac{dJ_0(\zeta)}{d\zeta} = -J_1(\zeta) \tag{4.8.3}$$

As before, it is convenient to put Eq.4.8.2 into nondimensional form by scaling the oscillatory shear stress by the corresponding shear stress in Poiseuille flow, that is, by dividing through by τ_s as given in Eq.3.4.6, to get

$$\frac{\tau_\phi(t)}{\tau_s} = \frac{2}{\Lambda}\left(\frac{J_1(\Lambda)}{J_0(\Lambda)}\right)e^{i\omega t} \tag{4.8.4}$$

The expression on the right-hand side is complex, its real part representing the shear stress at the tube wall when the driving pressure gradient varies as $\cos\omega t$ and its imaginary part representing that shear when the gradient varies as $\sin\omega t$. In both cases the result is numerically scaled by the corresponding shear stress in Poiseuille flow.

Variation of the imaginary part of $\tau_\phi(t)$ within the oscillatory cycle is shown in Fig.4.8.1. It is seen that it has a sinusoidal form like that of the imaginary part of the pressure gradient driving the flow, but with a phase difference between the two. The oscillatory shear stress lags the pressure, somewhat like the oscillatory flow rate. The amplitude of the oscillatory shear indicates the highest shear stress reached at the peak of each cycle as the fluid moves back and forth in each direction. This maximum clearly depends heavily on the frequency of oscillation as evident from Eq.4.8.4. The results shown in Fig.4.8.1 are for $\Omega = 3.0$, where it is seen that the oscillatory shear reaches a maximum value at the peak of each cycle of approximately one half the steady Poiseuille flow value.

In pulsatile flow consisting of steady- and oscillatory-flow components, the oscillatory shear stress adds to and subtracts from the steady shear stress. The results in Fig.4.8.1 show that in pulsatile flow at this particular frequency the shear stress would oscillate between a high of approximately 1.5 to a low of approximately 0.5 times its constant value in steady Poiseuille flow.

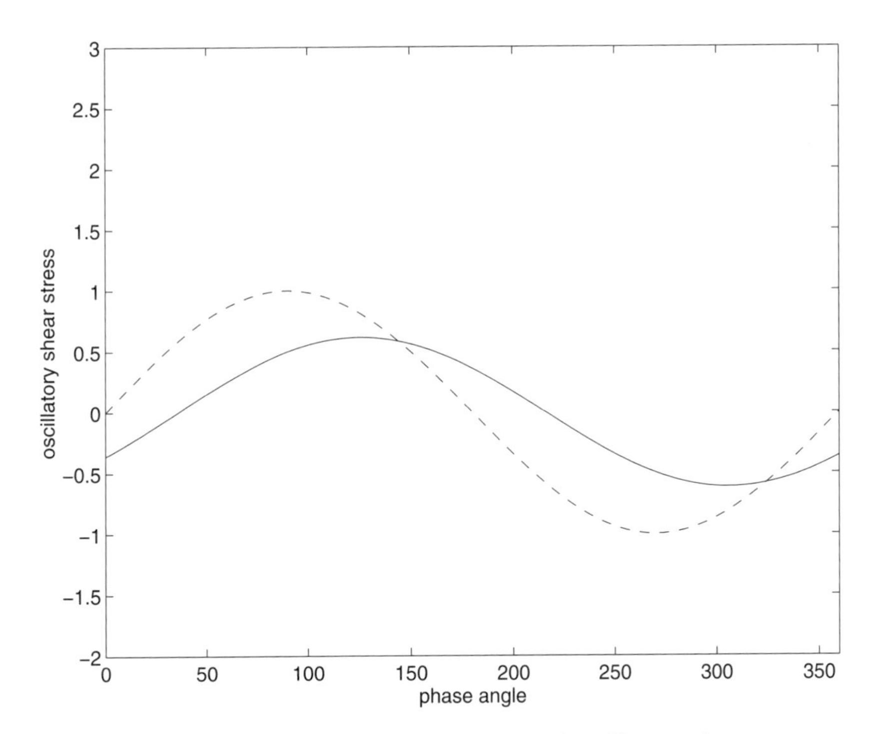

FIGURE 4.8.1. Variation of the imaginary part of oscillatory shear stress $\tau_{\phi I}$ (solid line) compared with the corresponding part of the pressure gradient $k_{\phi I}$ (dashed line). Shear stress *lags* the pressure, and its normalized value at its peak is less than 1.0, hence less than the corresponding shear stress in Poiseuille flow. Because the shear stress shown here is produced entirely by the oscillatory flow, it is *in addition* to the constant shear stress produced by the steady component of pulsatile flow.

4.9 Pumping Power

As in the case of steady Poiseuille flow, in pulsatile flow the equation governing the flow can be used to examine the balance of energy expenditure and in particular, in this case, to determine the pumping power required to drive the flow. Since the oscillatory part of pulsatile flow does not produce any net flow forward, and since the pumping power required to drive the steady part of pulsatile flow is the same as that in steady flow, any power expenditure on the oscillatory part of the flow is an "added expense" that reduces the efficiency of the flow. In this section we examine this added expense by considering the balance of energy expenditures in oscillatory flow, much along the same lines as was done for steady flow in Section 3.5.

We begin with the governing equation for oscillatory flow, namely, Eq.4.2.7

$$\mu \left(\frac{\partial^2 u_\phi}{\partial r^2} + \frac{1}{r} \frac{\partial u_\phi}{\partial r} \right) - \rho \frac{\partial u_\phi}{\partial t} = k_\phi(t)$$

For the purpose of present discussion we recall that in this equation both the velocity and pressure are complex quantities and that, therefore, the equation actually governs two separate problems, one for the real parts and one for the imaginary parts of the velocity and pressure. Using the notation of Section 4.6, the two governing equations are

$$\mu \left(\frac{\partial^2 u_{\phi R}}{\partial r^2} + \frac{1}{r} \frac{\partial u_{\phi R}}{\partial r} \right) - \rho \frac{\partial u_{\phi R}}{\partial t} = k_{\phi R} \qquad (4.9.1)$$

$$\mu \left(\frac{\partial^2 u_{\phi I}}{\partial r^2} + \frac{1}{r} \frac{\partial u_{\phi I}}{\partial r} \right) - \rho \frac{\partial u_{\phi I}}{\partial t} = k_{\phi I} \qquad (4.9.2)$$

Discussion of energy expenditure is more meaningful in terms of one or the other of these two equations, rather than in terms of the equation in complex form (Eq.4.2.7). Accordingly, the present discussion shall be based on Eq.4.9.2 that corresponds to an oscillatory pressure gradient varying as $\sin \omega t$, which was used in other sections in the present chapter. Similar discussion with obvious modifications can be based on Eq.4.9.1.

As in the case of steady Poiseuille flow, Eq.4.9.2 as it stands represents a balance of *forces per unit volume.* In the steady case this balance is between only the driving pressure term on the right-hand side and the viscous resistance term on the left-hand side. In the present case, however, there is an added acceleration term on the left-hand side. At any point in time, in oscillatory flow, the driving pressure force must equal the *net* sum of viscous and acceleration forces that may add to or subtract from each other at different times within the oscillatory cycle.

As in Section 3.5, we consider a specific volume of fluid consisting of a cylindrical shell of radius r, length l, and thickness dr, moving with velocity $u_{\phi I}(r, t)$. It is important to recall that axial location x along the tube is not a factor in pulsatile flow through a rigid tube because fluid at all cross sections of the tube is moving with the same velocity profile, hence x does not appear as a variable in the velocity or in the governing equation.

If each of the three terms in Eq.4.9.2 is multiplied by the volume of this cylindrical shell of fluid, namely, $2\pi r l dr$, and by the velocity $u_{\phi I}$, the result is an equation governing the balance of energy expenditures associated with this volume of fluid. Writing

$$dH_{vI} = \mu \left(\frac{\partial^2 u_{\phi I}}{\partial r^2} + \frac{1}{r} \frac{\partial u_{\phi I}}{\partial r} \right) \times 2\pi r l u_{\phi I} dr \qquad (4.9.3)$$

$$dH_{aI} = \rho \frac{\partial u_{\phi I}}{\partial t} \times 2\pi r l u_{\phi I} dr \qquad (4.9.4)$$

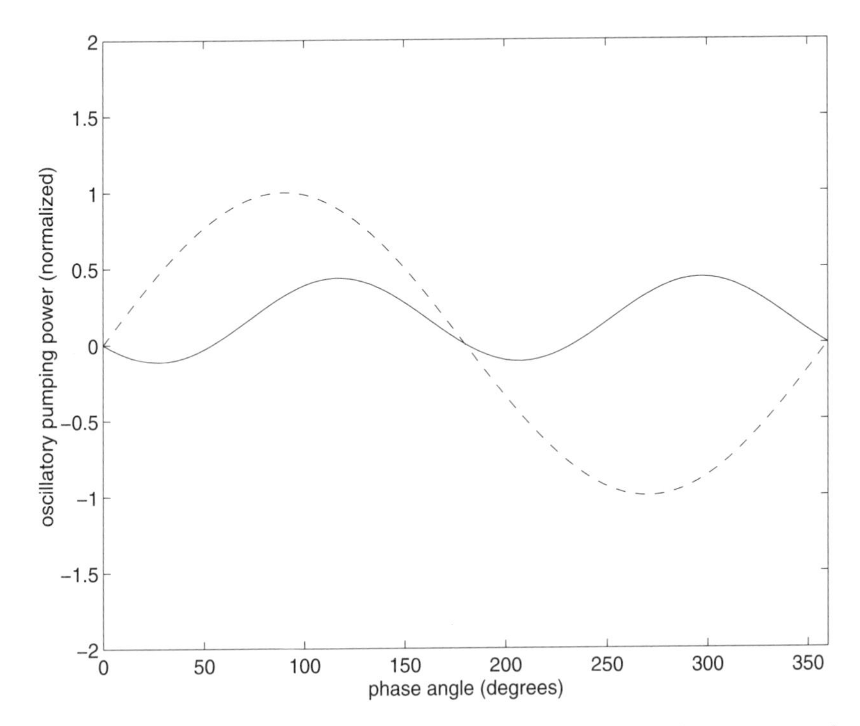

FIGURE 4.9.1. Variation of oscillatory pumping power $H_{\phi I}$ during one cycle (solid line) compared with the corresponding pressure gradient $k_{\phi I}$ (dashed line). The power has two peaks within each oscillatory cycle because it consists of the product of oscillatory pressure and oscillatory flow. The integral of the power over one cycle is not zero, hence oscillatory flow requires energy to maintain even though the net flow is zero. This energy expenditure is required to maintain the energy dissipation at the tube wall. The net energy expenditure for accelerating and decelerating the flow is zero (see text).

$$dH_{pI} = k_{\phi I} \times 2\pi r l u_{\phi I} dr \tag{4.9.5}$$

Eq.4.9.2 becomes

$$dH_{vI}(r,t) - dH_{aI}(r,t) = dH_{pI}(r,t) \tag{4.9.6}$$

The notation emphasizes the fact that each term is a function of r and t, and that the imaginary parts of the velocity and pressure are being used for illustration. Subscripts v, a, p are used to associate energy expenditure with viscous dissipation, acceleration, and pressure, respectively. If now each term is integrated over a cross section of the tube as was done in the steady flow case, the r dependence is removed and the equation then becomes a balance of energy expenditure for a cylindrical volume of fluid of radius a and length l filling the entire cross section of the tube. If the results of integration are denoted by H_v, H_a, H_p, respectively, the equation becomes

$$H_{vI}(t) - H_{aI}(t) = H_{pI}(t) \tag{4.9.7}$$

where, for each term,

$$H(t) = \int_{r=0}^{r=a} dH(r,t) \tag{4.9.8}$$

Eq.4.9.7 expresses the balance of energy expenditures *at each point in time* within the oscillatory cycle.

If each term in Eq.4.9.7 is now integrated in time over one cycle of period $T = 2\pi/\omega$, and if we denote the results by E_v, E_a, E_p, respectively, then the equation becomes

$$E_v - E_a = E_p \tag{4.9.9}$$

where, for each term,

$$E = \int_0^T H(t)dt \tag{4.9.10}$$

Eq.4.9.9 expresses the balance of energy expenditures *over one complete oscillatory cycle*, a real/imaginary subscript is not required since the equation holds in both cases.

At a particular point in time within the oscillatory cycle, the viscous component of the energy expenditure, using Eq.4.9.3, is given by

$$\begin{aligned}
H_{vI}(t) &= \int_{r=0}^{r=a} dH_{vI}(r,t) \\
&= 2\pi\mu l \int_0^a u_{\phi I}\left(\frac{\partial^2 u_{\phi I}}{\partial r^2} + \frac{1}{r}\frac{\partial u_{\phi I}}{\partial r}\right) rdr \\
&= 2\pi\mu l \int_{r=0}^{r=a} u_{\phi I}\frac{\partial}{\partial r}\left(r\frac{\partial u_{\phi I}}{\partial r}\right) dr \\
&= 2\pi\mu l \left| u_{\phi I}\frac{\partial u_{\phi I}}{\partial r} r\right|_{r=0}^{r=a} - \int_{r=0}^{r=a} \frac{\partial u_{\phi I}}{\partial r} r du_{\phi I} \\
&= -2\pi\mu l \int_0^a \left(\frac{\partial u_{\phi I}}{\partial r}\right)^2 rdr
\end{aligned} \tag{4.9.11}$$

which, as in the case of steady flow, represents the rate of energy expenditure due to viscous dissipation. And because this dissipation is due to the oscillatory velocity component $u_{\phi I}$, this energy expenditure is due entirely to the oscillatory component of the flow.

The accelerating component of energy expenditure, using Eq.4.9.4, is given by

$$H_{aI}(t) = \int_{r=0}^{r=a} dH_{aI}(r,t)$$

$$
\begin{aligned}
&= 2\pi\rho l \int_0^a u_{\phi I}\left(\frac{\partial u_{\phi I}}{\partial t}\right) r dr \\
&= 2\pi\rho l \int_0^a \frac{\partial}{\partial t}\left(\frac{u_{\phi I}^2}{2}\right) r dr \\
&= 2\pi l \frac{d}{dt}\int_0^a \left(\frac{1}{2}\rho u_{\phi I}^2\right) r dr \qquad (4.9.12)
\end{aligned}
$$

which represents the rate of energy expenditure required for accelerating the flow, increasing its kinetic energy as indicated under the integral sign.

The driving (pressure) component of energy expenditure, using Eq.4.9.5, is given by

$$
\begin{aligned}
H_{pI}(t) &= \int_{r=0}^{r=a} dH_{pI}(t) \\
&= lk_{\phi I}\int_0^a 2\pi r u_{\phi I} dr \\
&= lk_{\phi I} q_{\phi I} \qquad (4.9.13)
\end{aligned}
$$

which represents the pumping power required to drive the flow and where $q_{\phi I}$ is the imaginary part of the flow rate, from Eq.4.7.5

$$
q_{\phi I} = \Im\left\{\frac{i\pi k_s a^4}{\mu\Omega^2}\left(1-\frac{2J_1(\Lambda)}{\Lambda J_0(\Lambda)}\right)e^{i\omega t}\right\} \qquad (4.9.14)
$$

Thus the *instantaneous* balance of energy expenditure, by substituting into Eq.4.9.7 from Eqs.4.9.11–13, can be written as

$$
-2\pi\mu l \int_0^a r\left(\frac{\partial u_{\phi I}}{\partial r}\right)^2 dr + 2\pi l \frac{d}{dt}\int_0^a \left(\frac{1}{2}\rho u_{\phi I}^2\right) r dr = lk_{\phi I} q_{\phi I} \qquad (4.9.15)
$$

from which it is seen that, at any point in time, the pumping power on the right-hand side is being expended on the net sum of energy required to accelerate the flow and that required to overcome the viscous resistance. Since as the oscillatory cycle progresses the acceleration term changes sign while the viscous term does not, the net result at any instant may represent the sum or difference of the two terms. Physically, this means that during the acceleration phase of the oscillatory cycle the pumping power pays for both acceleration and viscous dissipation, while during the deceleration phase the flow actually returns some of its kinetic energy.

It should be noted that if the pumping power is to be defined as a *positive* quantity as was done for steady flow (Eq.3.4.14), then we introduce

$$
H_{\phi I}(t) = -lk_{\phi I} q_{\phi I} = -H_{pI} \qquad (4.9.16)
$$

$$H_{\phi R}(t) = -lk_{\phi R}q_{\phi R} = -H_{pR} \tag{4.9.17}$$

Variation of oscillatory pumping power is shown in Fig.4.9.1, where it is seen that the power has two peaks within one oscillatory cycle because it consists of the product of k_ϕ and q_ϕ. The figure also shows clearly that the integral of the power over one cycle is not zero.

By contrast, the energy expenditure for accelerating and decelerating the flow over one cycle, if $T = 2\pi/\omega$ is the period of oscillation, is given by

$$\begin{aligned}
E_a &= \int_0^T e_a(t)dt \\
&= 2\pi\rho l \int_0^{2\pi/\omega}\int_0^a u_{\phi I}\frac{\partial u_{\phi I}}{\partial t} rdrdt \\
&= 2\pi\rho l \int_0^a\int_0^{2\pi/\omega} u_{\phi I}\frac{\partial u_{\phi I}}{\partial t} dtrdr \\
&= 2\pi\rho l \int_0^a \left\{\int_{t=0}^{t=2\pi/\omega} d\left(\frac{u_{\phi I}^2}{2}\right)\right\} rdr \\
&= 2\pi\rho l \int_0^a \left|\frac{u_{\phi I}^2}{2}\right|_{t=0}^{t=2\pi/\omega} rdr \\
&= 2\pi\rho l \int_0^a \left|\frac{(U_{\phi I}\cos\omega t + U_{\phi R}\sin\omega t)^2}{2}\right|_{t=0}^{t=2\pi/\omega} rdr \\
&= 0
\end{aligned} \tag{4.9.18}$$

Thus while the instantaneous energy expenditure $H_a(t)$ required to accelerate the flow is generally nonzero, the net expenditure over one cycle is zero. The energy spent during one half of the cycle is recovered during the other half.

It follows, therefore, that *over one cycle* the average power expenditure required to drive the oscillatory part of the flow is equal to that being dissipated by viscosity, that is,

$$E_p = E_v \tag{4.9.19}$$

The *positive* average rate of energy expenditure over one cycle is thus given by

$$\frac{E_\phi}{2\pi/\omega} = \frac{1}{2\pi/\omega}\int_0^{2\pi/\omega} H_{\phi I}(t)dt = \frac{1}{2\pi/\omega}\int_0^{2\pi/\omega} H_{\phi R}(t)dt \tag{4.9.20}$$

$$= -\frac{l}{2\pi/\omega}\int_0^{2\pi/\omega} k_{\phi I} q_{\phi I}(t)dt \qquad (4.9.21)$$

$$= -\frac{l}{2\pi/\omega}\int_0^{2\pi/\omega} k_{\phi R} q_{\phi R}(t)dt \qquad (4.9.22)$$

and expressing this as a fraction of the corresponding pumping power in steady flow (Eq.3.4.14), we get

$$\frac{E_\phi}{H_s \times 2\pi/\omega} = \frac{1}{2\pi/\omega}\int_0^{2\pi/\omega} \left(\frac{k_{\phi I}}{k_s}\right)\left(\frac{q_{\phi I}}{q_s}\right) dt \qquad (4.9.23)$$

$$= \frac{1}{2\pi/\omega}\int_0^{2\pi/\omega} \left(\frac{k_{\phi R}}{k_s}\right)\left(\frac{q_{\phi R}}{q_s}\right) dt \qquad (4.9.24)$$

These results are used in the next section to assess the magnitude of this energy expenditure at high and at low frequency.

4.10 Oscillatory Flow at Low Frequency

At low frequency oscillatory flow in a tube is better able to keep pace with the changing pressure. In fact at very low frequency, or in the limit of "zero frequency," the relation between flow and pressure becomes *instantaneously* the same as in steady Poiseuille flow. That is, at each point in time within the oscillatory cycle the velocity profile is what it would be in steady Poiseuille flow under a pressure gradient equal to the value of the oscillatory pressure gradient at that instant. The situation suggests the term "oscillatory Poiseuille flow." In this section we demonstrate these features of the flow analytically and derive approximate expressions which are valid at low frequency and which are easier to use than the more general expressions involving Bessel functions.

At a frequency of 1 *cycle per second*, which is equivalent to an angular frequency of 2π *radians per second*, density of 1 gm/cm^3, and viscosity of 0.04 *Poise* (*Poise* = *dyne second*/cm^2), the value of the nondimensional frequency parameter Ω, using Eq.4.4.5, is given by

$$\Omega = \sqrt{\frac{2\pi}{0.04}}\, a \qquad (4.10.1)$$

where a is the radius of the tube in cm. Thus for a tube of 1 cm radius, the value of Ω is approximately 12.5. In the human system, therefore, $\Omega = 1$ may be taken as a representative low value of the frequency parameter and $\Omega = 10$ as a moderately high value.

A series expansion of the Bessel function $J_0(z)$ for small z is given by [4,5]

$$J_0(z) = 1 - \frac{z^2}{2^2} + \frac{z^4}{2^2 \times 4^2} - \frac{z^6}{2^2 \times 4^2 \times 6^2} + \ldots \tag{4.10.2}$$

The expansion is valid for *complex* values of the independent variable z as required for application to the complex solution of the pulsatile flow equation. Thus an approximation of the quotient term in Eq.4.6.2 for the velocity profile, using only the first three terms of the series, and recalling from Eq.4.5.4 that $\zeta = \Lambda r/a$, is given by

$$\begin{aligned}
\frac{J_0(\zeta)}{J_0(\Lambda)} &= \frac{J_0(\Lambda r/a)}{J_0(\Lambda)} \\
&\approx \left(1 - \frac{\Lambda^2 r^2}{4a^2} + \frac{\Lambda^4 r^4}{64a^4}\right) \times \left(1 - \frac{\Lambda^2}{4} + \frac{\Lambda^4}{64}\right)^{-1} \\
&\approx \left(1 - \frac{\Lambda^2 r^2}{4a^2} + \frac{\Lambda^4 r^4}{64a^4}\right) \times \left(1 + \frac{\Lambda^2}{4} + \frac{3\Lambda^4}{64}\right) \\
&\approx 1 + \frac{\Lambda^2}{4}\left\{\left(1 - \frac{r^2}{a^2}\right) + \frac{\Lambda^2}{16}\left(3 - \frac{4r^2}{a^2} + \frac{r^4}{a^4}\right)\right\}
\end{aligned} \tag{4.10.3}$$

only terms of order ζ^4 being retained at each step. By substituting this result into Eq.4.6.2 for the velocity profile, we obtain the following approximate expression:

$$\frac{u_\phi(r,t)}{\hat{u}_s} \approx \left\{\left(1 - \frac{r^2}{a^2}\right) - \frac{i\Omega^2}{16}\left(3 - \frac{4r^2}{a^2} + \frac{r^4}{a^4}\right)\right\} e^{i\omega t} \tag{4.10.4}$$

The real and imaginary parts of the velocity are then given by

$$\frac{u_{\phi R}(r,t)}{\hat{u}_s} \approx \left(1 - \frac{r^2}{a^2}\right)\cos\omega t + \frac{\Omega^2}{16}\left(3 - \frac{4r^2}{a^2} + \frac{r^4}{a^4}\right)\sin\omega t \tag{4.10.5}$$

$$\frac{u_{\phi I}(r,t)}{\hat{u}_s} \approx \left(1 - \frac{r^2}{a^2}\right)\sin\omega t - \frac{\Omega^2}{16}\left(3 - \frac{4r^2}{a^2} + \frac{r^4}{a^4}\right)\cos\omega t \tag{4.10.6}$$

These expressions are easier to use than Eq.4.6.2 because they do not involve Bessel functions and can be used in place of that equation when the frequency is small. Furthermore, substituting for $\hat{u}_s$ from Eq.3.4.1, and using Eq.4.6.5 for the real and imaginary parts of the pressure gradient, we obtain

$$u_{\phi R}(r,t) \approx -\frac{k_{\phi R}a^2}{4\mu}\left\{\left(1 - \frac{r^2}{a^2}\right) + \frac{\Omega^2}{16}\left(3 - \frac{4r^2}{a^2} + \frac{r^4}{a^4}\right)\tan\omega t\right\} \tag{4.10.7}$$

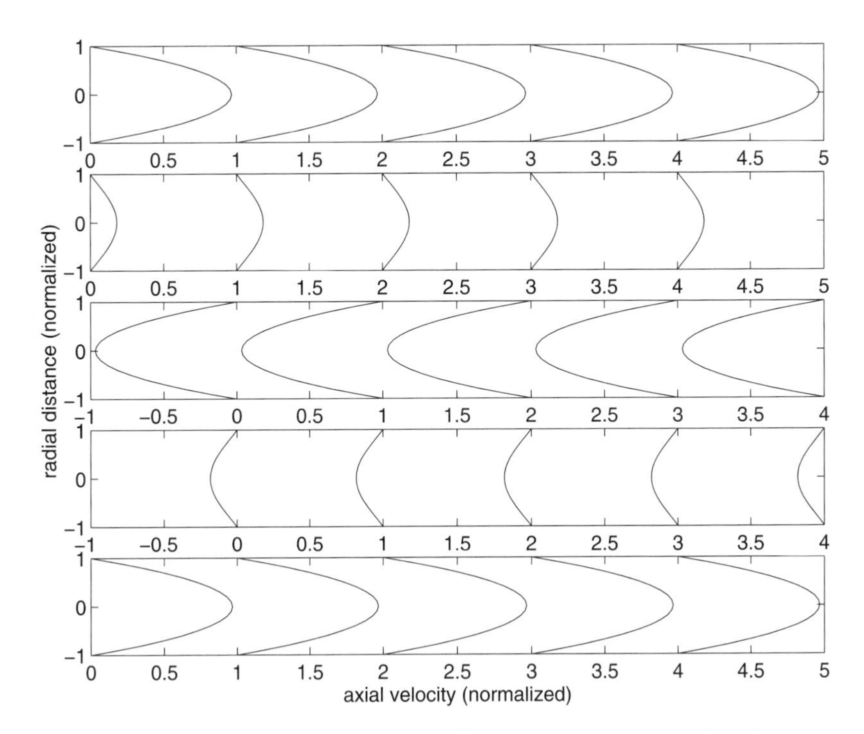

FIGURE 4.10.1. Oscillatory velocity profiles in a rigid tube at low frequency ($\Omega = 1.0$) and corresponding to the *real* part of the pressure gradient, namely, $k_s \cos\omega t$. The panels represent the profiles at different phase angles (ωt) within the oscillatory cycle, with $\omega t = 0$ in the top panel then increasing by 90° for subsequent panels. The profiles reach their peak form at the peak of pressure gradient ($\omega t = 0°, 180°$), and the maximum velocity at peak has a normalized value near 1.0. Thus flow is almost in phase with pressure gradient, and the relation between the two is as if the flow were Poiseuille flow *at each instant* (see text).

$$u_{\phi I}(r,t) \approx -\frac{k_{\phi I}a^2}{4\mu}\left\{\left(1-\frac{r^2}{a^2}\right) - \frac{\Omega^2}{16}\left(3-\frac{4r^2}{a^2}+\frac{r^4}{a^4}\right)\cot\omega t\right\} \tag{4.10.8}$$

We see that for small Ω where the second term in each of the two expressions can be neglected, the relation between velocity and pressure becomes

$$u_{\phi R}(r,t) \approx \frac{k_{\phi R}}{4\mu}(r^2 - a^2) \tag{4.10.9}$$

$$u_{\phi I}(r,t) \approx \frac{k_{\phi I}}{4\mu}(r^2 - a^2) \tag{4.10.10}$$

which is the same as that in Eq.3.3.10 for steady Poiseuille flow, but with instantaneous values of velocity and pressure, hence justifying the term

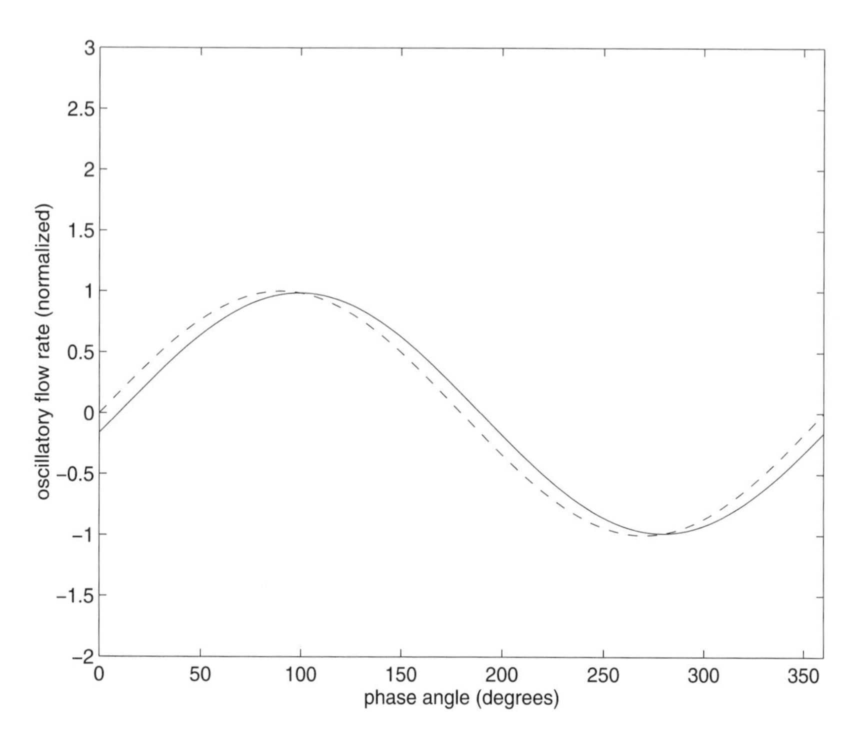

FIGURE 4.10.2. Variation of the oscillatory flow rate q_ϕ within an oscillatory cycle (solid line), compared with variation of the corresponding pressure gradient, in this case $k_{\phi I} = k_s \sin \omega t$ (dashed line), at low frequency ($\Omega = 1.0$). Flow is almost in phase with pressure gradient, and normalized peak flow is close to 1.0. Flow at each point in time is close to what it would be in steady Poiseuille flow under the instantaneous value of the pressure gradient.

"oscillatory Poiseuille flow." Velocity profiles with $\Omega = 1.0$ are shown in Fig.4.10.1.

For the flow rate we use the series expansion of $J_1(\zeta)$ for small values of ζ [4,5]:

$$J_1(\zeta) = \frac{\zeta}{2} - \frac{\zeta^3}{2^2 \times 4} + \frac{\zeta^5}{2^2 \times 4^2 \times 6} + \frac{\zeta^7}{2^2 \times 4^2 \times 6^2 \times 8} + \dots \tag{4.10.11}$$

Using only the first three terms of the series for the quotient term in Eq.4.7.7, we find

$$\begin{aligned}\frac{J_1(\Lambda)}{J_0(\Lambda)} &\approx \frac{\Lambda}{2}\left(1 - \frac{\Lambda^2}{8} + \frac{\Lambda^4}{192}\right) \times \left(1 - \frac{\Lambda^2}{4} + \frac{\Lambda^4}{64}\right)^{-1} \\ &\approx \frac{\Lambda}{2}\left(1 - \frac{\Lambda^2}{8} + \frac{\Lambda^4}{192}\right) \times \left(1 + \frac{\Lambda^2}{4} + \frac{3\Lambda^4}{64}\right)\end{aligned}$$

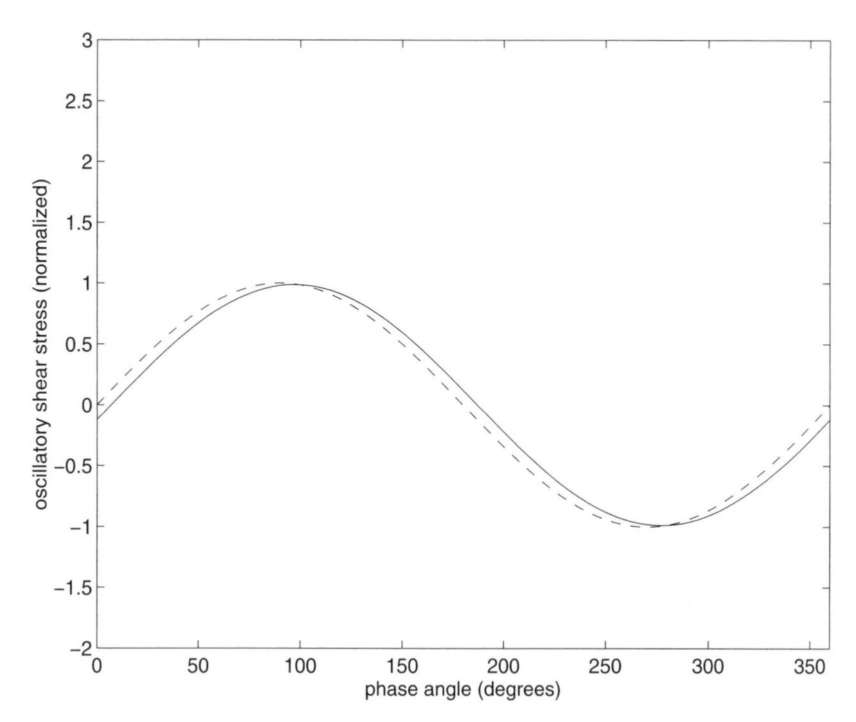

FIGURE 4.10.3. Variation of the imaginary part of oscillatory shear stress $\tau_{\phi I}$ (solid line) compared with the corresponding part of the pressure gradient $k_{\phi I}$ (dashed line), at low frequency ($\Omega = 1.0$). Shear stress is almost in phase with pressure gradient, and normalized peak shear stress is close to 1.0. Shear stress at each point in time is close to what it would be in steady Poiseuille flow under the instantaneous value of the pressure gradient.

$$\approx \frac{\Lambda}{2}\left(1+\frac{\Lambda^2}{8}+\frac{\Lambda^4}{48}\right) \tag{4.10.12}$$

where again, only terms of order Λ^4 or lower were retained at each step. Substituting this result into Eq.4.7.7, we get

$$\frac{q_\phi(t)}{q_s} \approx \frac{8i}{\Omega^2}\left(\frac{\Lambda^2}{8}+\frac{\Lambda^4}{48}\right) e^{i\omega t} \tag{4.10.13}$$

Noting that $\Lambda^2 = -i\Omega^2$ and $\Lambda^4 = -\Omega^4$, the real and imaginary parts of the flow rate are given by

$$\frac{q_{\phi R}(t)}{q_s} \approx \cos\omega t + \frac{\Omega^2}{6}\sin\omega t \tag{4.10.14}$$

$$\frac{q_{\phi I}(t)}{q_s} \approx \sin\omega t - \frac{\Omega^2}{6}\cos\omega t \tag{4.10.15}$$

Substituting for the steady-flow rate from Eq.3.4.3, and for the real and imaginary parts of the pressure gradient from Eq.4.6.5, we get

$$q_{\phi R} \approx \frac{-k_{\phi R}\pi a^4}{8\mu}\left(1+\frac{\Omega^2}{6}\tan\omega t\right) \tag{4.10.16}$$

$$q_{\phi I} \approx \frac{-k_{\phi I}\pi a^4}{8\mu}\left(1-\frac{\Omega^2}{6}\cot\omega t\right) \tag{4.10.17}$$

At low frequency where the second term in each expression can be neglected, the relation between flow rate and pressure gradient becomes the same as that in steady flow *at each instant*, that is,

$$q_{\phi R} \approx \frac{-k_{\phi R}\pi a^4}{8\mu} \tag{4.10.18}$$

$$q_{\phi I} \approx \frac{-k_{\phi I}\pi a^4}{8\mu} \tag{4.10.19}$$

as in Eq.3.4.3 for steady flow. Flow rate with $\Omega = 1.0$ is shown in Fig.4.10.2.

In the same way, and omitting the details, we obtain the following expressions for the shear stress and maximum velocity:

$$\tau_{\phi R} \approx -\frac{k_{\phi R}a}{2}\left(1+\frac{\Omega^2}{8}\tan\omega t\right) \tag{4.10.20}$$

$$\tau_{\phi I} \approx -\frac{k_{\phi I}a}{2}\left(1-\frac{\Omega^2}{8}\cot\omega t\right) \tag{4.10.21}$$

$$\hat{u}_{\phi R} \approx \frac{-k_{\phi R}a^2}{4\mu}\left(1+\frac{3\Omega^2}{16}\tan\omega t\right) \tag{4.10.22}$$

$$\hat{u}_{\phi I} \approx \frac{-k_{\phi I}a^2}{4\mu}\left(1-\frac{3\Omega^2}{16}\cot\omega t\right) \tag{4.10.23}$$

In each case, if the frequency is low enough for the second term to be neglected, the relation becomes *instantaneously* the same as in steady flow (Eqs.3.4.1,6). If the frequency is small but not negligible, the second term can be used for approximate calculations. Oscillatory shear stress with $\Omega = 1.0$ is shown in Fig.4.10.3.

For the pumping power we have from Eqs.4.9.16,17 and the approximate results for the flow rate in Eqs.4.10.16,17

$$\begin{aligned} H_{\phi R}(t) &= -lk_{\phi R}q_{\phi R} \\ &\approx \frac{\pi a^4 l}{8\mu}k_s^2\left(\cos^2\omega t+\frac{\Omega^2}{6}\cos\omega t\sin\omega t\right) \end{aligned} \tag{4.10.24}$$

$$H_{\phi I}(t) = -lk_{\phi I}q_{\phi I}$$

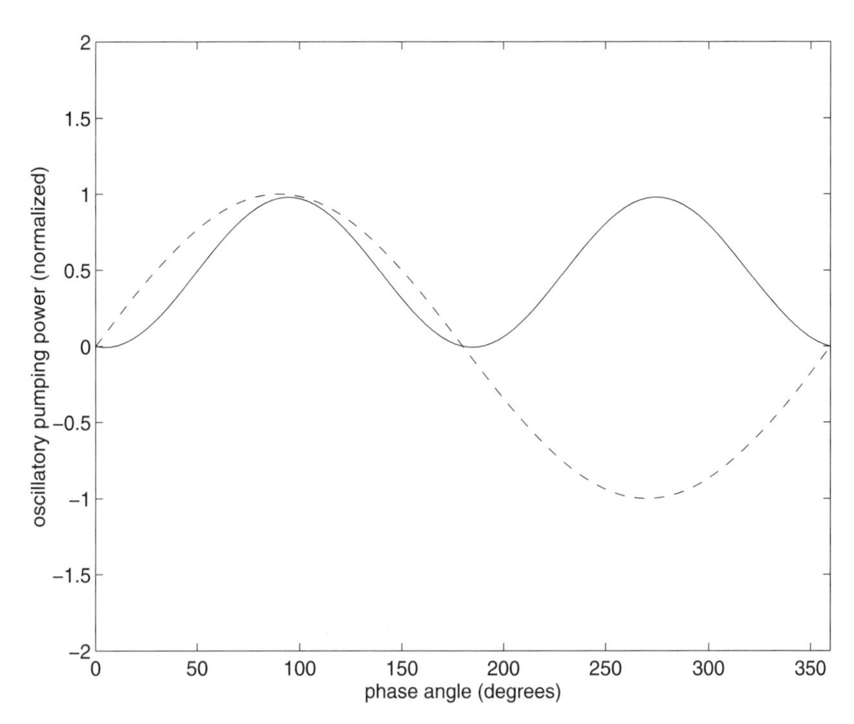

FIGURE 4.10.4. Variation of oscillatory pumping power $H_{\phi I}$ during one cycle (solid line) compared with the corresponding pressure gradient $k_{\phi I}$ (dashed line), at low frequency ($\Omega = 1.0$). The two peaks of the power coincide very closely with the peaks of pressure gradient, which at low frequency coincide with the peaks of flow rate (Fig.4.10.2). The area under the power curve, which represents the net energy expenditure over one cycle, is not zero. In fact, the area equals one half the corresponding energy expenditure in steady Poiseuille flow (see text).

$$\approx \quad \frac{\pi a^4 l}{8\mu} k_s^2 \left(\sin^2 \omega t - \frac{\Omega^2}{6} \cos \omega t \sin \omega t \right) \tag{4.10.25}$$

depending on whether the driving pressure gradient is the real or imaginary part of k_ϕ, respectively.

It is interesting to note that the sum of the two results is a constant independent of time and equal to the pumping power required in steady Poiseuille flow. Because the energy expenditure over one cycle is the same whether the driving pressure gradient is $k_{\phi R}$ or $k_{\phi I}$, this implies that the *average pumping power* in oscillatory flow at low frequency is one half the corresponding power in steady flow. This can be easily verified by noting that

$$\int_0^{2\pi} \cos^2 \omega t d(\omega t) = \int_0^{2\pi} \sin^2 \omega t d(\omega t) = \pi \tag{4.10.26}$$

while

$$\int_0^{2\pi} \cos\omega t \sin\omega t d(\omega t) = 0 \tag{4.10.27}$$

Thus whether the flow is being driven by the real or imaginary part of the pressure gradient, using Eqs.4.9.20,23,24, the average pumping power required for the oscillatory part of the flow, as a fraction of the corresponding power in steady flow, is given by

$$\frac{E_\phi}{H_s \times 2\pi/\omega} \approx \frac{1}{2\pi/\omega}\int_0^{2\pi/\omega} \sin^2\omega t dt \tag{4.10.28}$$

$$\approx \frac{1}{2\pi/\omega}\int_0^{2\pi/\omega} \cos^2\omega t dt \tag{4.10.29}$$

$$\approx \frac{1}{2} \tag{4.10.30}$$

Since in oscillatory flow there is no net flow forward, this pumping power is "wasted" in the sense that it is not being utilized to a useful end. Thus in pulsatile flow, at low frequency, the total power required to drive the flow is equal to the power required to drive the steady part of the flow, as in Poiseuille flow, plus one half of that amount to maintain the oscillation of the oscillatory part of the flow. Pumping power with $\Omega = 1.0$ is shown in Fig.4.10.4.

4.11 Oscillatory Flow at High Frequency

At high frequency oscillatory flow in a tube is less able to keep pace with the changing pressure, thus reaching less than the fully developed Poiseuille flow profile at the peak of each cycle. The higher the frequency, the lower the peak velocity the flow is able to reach. In the limit of infinite frequency the velocity reached at the peak of each cycle is zero, that is, the fluid does not move at all. An interesting question is whether the pumping power required to maintain this limiting state of zero flow is zero. In this section we develop approximate expressions describing properties of oscillatory flow at high frequency, which are easier to use than the more general expressions involving Bessel functions and which will be used to answer this question.

An approximate expression for $J_0(\zeta)$ when ζ is large is given by [4,5]

$$J_0(\zeta) \approx \frac{\sin\zeta + \cos\zeta}{\sqrt{\pi\zeta}} \tag{4.11.1}$$

For the purpose of algebraic manipulation we write $\zeta = i\zeta_1$ so that

$$J_0(\zeta) \approx \frac{\sin(i\zeta_1) + \cos(i\zeta_1)}{\sqrt{\pi i\zeta_1}}$$

$$\begin{aligned}
&\approx \frac{i\sinh\zeta_1+\cosh\zeta_1}{\sqrt{\pi i\zeta_1}} \\
&\approx \frac{(1+i)}{2}\frac{e^{\zeta_1}}{\sqrt{\pi i\zeta_1}}
\end{aligned} \tag{4.11.2}$$

and similarly, writing $\Lambda = i\Lambda_1$,

$$J_0(\Lambda) \approx \frac{(1+i)}{2}\frac{e^{\Lambda_1}}{\sqrt{\pi i\Lambda_1}} \tag{4.11.3}$$

Near the tube wall where $r/a \approx 1$: Inserting the above approximations into Eq.4.6.2 for the velocity profile, and recalling from Eq.4.5.4 that $\zeta = \Lambda r/a$, hence $\zeta_1 = \Lambda_1 r/a$, we find

$$\begin{aligned}
\frac{u_\phi(r,t)}{\hat{u}_s} &= \frac{-4}{\Lambda^2}\left(1-\frac{J_0(\zeta)}{J_0(\Lambda)}\right)e^{i\omega t} \\
&\approx \frac{-4}{\Lambda^2}\left(1-\sqrt{\frac{\Lambda_1}{\zeta_1}}e^{(\zeta_1-\Lambda_1)}\right)e^{i\omega t} \\
&\approx \frac{-4}{\Lambda^2}\left(1-\sqrt{\frac{a}{r}}e^{\Lambda_1(\frac{r}{a}-1)}\right)e^{i\omega t} \\
&\approx \frac{4i}{\Lambda}\left(1-\frac{r}{a}\right)e^{i\omega t}
\end{aligned} \tag{4.11.4}$$

The real and imaginary parts of which are given by

$$\frac{u_{\phi R}(r,t)}{\hat{u}_s} \approx \frac{2\sqrt{2}}{\Omega}\left(1-\frac{r}{a}\right)(\cos\omega t+\sin\omega t) \tag{4.11.5}$$

$$\frac{u_{\phi I}(r,t)}{\hat{u}_s} \approx \frac{2\sqrt{2}}{\Omega}\left(1-\frac{r}{a}\right)(\sin\omega t-\cos\omega t) \tag{4.11.6}$$

Near the center of the tube, where $r/a \approx 0$ but Λ is large because of high frequency: Here the ratio $J_0(\zeta)/J_0(\Lambda)$ in the expression for the velocity (Eq.4.6.2) requires an approximation of $J_0(\zeta)$ for *small* ζ and an expansion of $J_0(\Lambda)$ for *large* Λ. The ratio ultimately vanishes, with the result

$$\frac{u_\phi(r,t)}{\hat{u}_s} = \frac{-4}{\Lambda^2}e^{i\omega t} \tag{4.11.7}$$

the real and imaginary parts of which are given by

$$\frac{u_{\phi R}(r,t)}{\hat{u}_s} \approx \frac{4}{\Omega^2}\sin\omega t \tag{4.11.8}$$

$$\frac{u_{\phi I}(r,t)}{\hat{u}_s} \approx \frac{-4}{\Omega^2}\cos\omega t \tag{4.11.9}$$

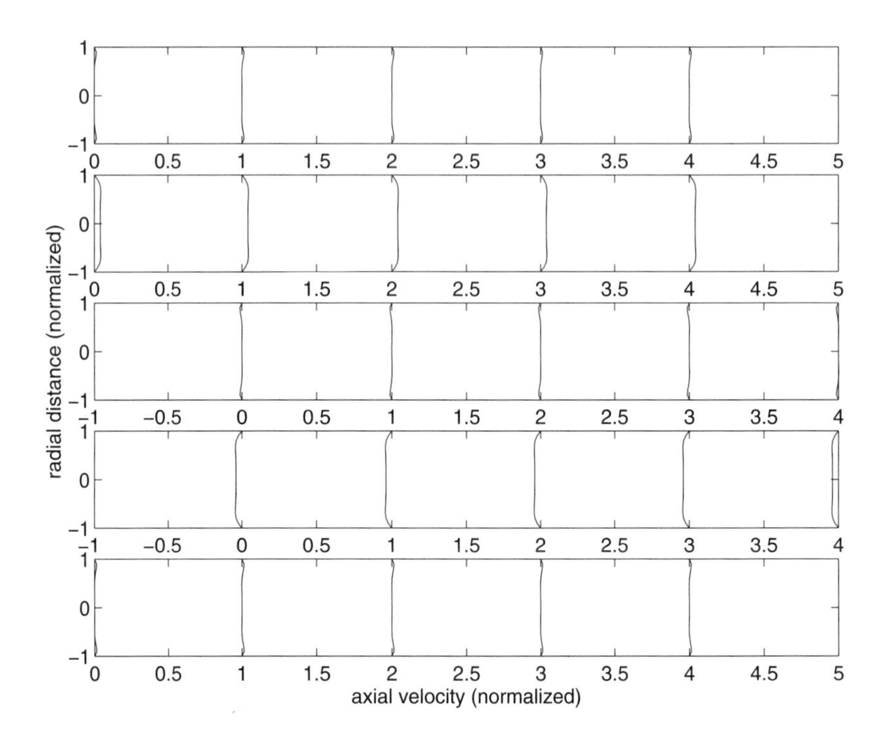

FIGURE 4.11.1. Oscillatory velocity profiles in a rigid tube at high frequency ($\Omega = 10$) and corresponding to the *real* part of the pressure gradient, namely, $k_s \cos \omega t$. The panels represent the profiles at different phase angles (ωt) within the oscillatory cycle, with $\omega t = 0$ in the top panel then increasing by 90° for subsequent panels. While the velocity is everywhere near zero, the profiles reach their peak form in the second panel, which means that they are about 90° out of phase with the pressure gradient (see text).

These results indicate that at high frequency fluid near the tube wall is affected differently from fluid near the center of the tube, thus distorting the parabolic character of the velocity profile. There is some phase difference between the velocity near the center of the tube and that near the wall. By contrast, at *low* frequency fluid is affected more uniformly by pulsation, and the parabolic character of the velocity profile is fairly well preserved during the oscillatory cycle, as we saw in the previous section. Velocity profiles with $\Omega = 10$ are shown in Fig.4.11.1.

For the flow rate, similarly, we use the approximation for large ζ of the Bessel function of first order, namely [4,5],

$$J_1(\zeta) \approx \frac{\sin \zeta - \cos \zeta}{\sqrt{\pi \zeta}} \tag{4.11.10}$$

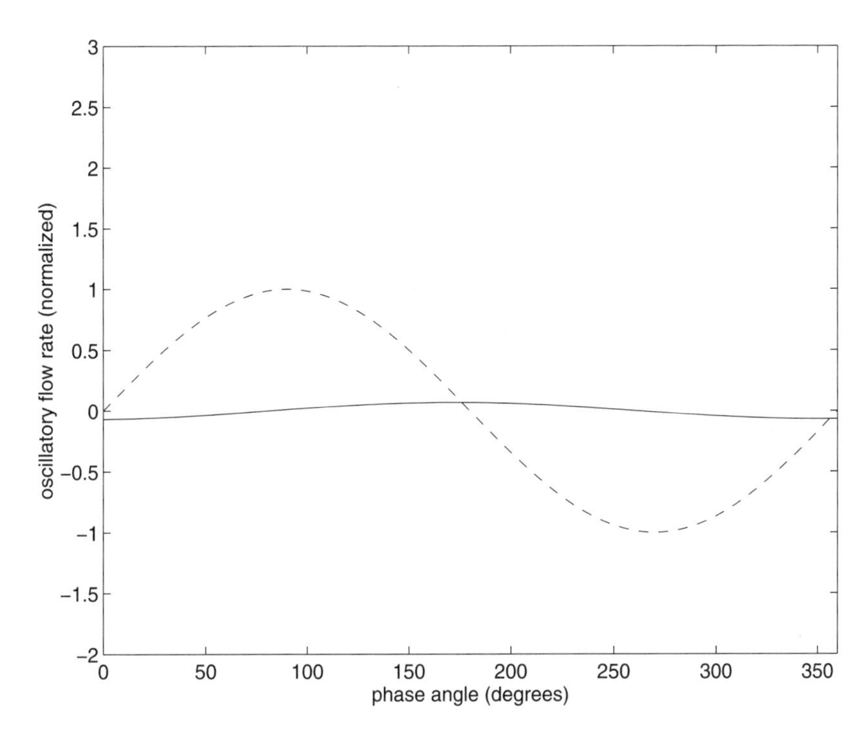

FIGURE 4.11.2. Variation of the oscillatory flow rate q_ϕ within an oscillatory cycle (solid line), compared with variation of the corresponding pressure gradient, in this case $k_{\phi I} = k_s \sin \omega t$ (dashed line), at high frequency ($\Omega = 10$). Flow rate is almost zero throughout the cycle, but there is an approximately 90° phase difference between flow rate and pressure gradient.

As before, writing $\zeta = i\zeta_1,\ \Lambda = i\Lambda_1$, gives

$$J_1(\zeta) \approx \frac{(i-1)e^{\zeta_1}}{2\sqrt{\pi i \zeta_1}} \tag{4.11.11}$$

$$J_1(\Lambda) \approx \frac{(i-1)e^{\Lambda_1}}{2\sqrt{\pi i \Lambda_1}} \tag{4.11.12}$$

and substituting into Eq.4.7.7 we get

$$\begin{aligned}
\frac{q_\phi(t)}{q_s} &= \frac{-8i}{\Omega^2}\left(1 - \frac{2J_1(\Lambda)}{\Lambda J_0(\Lambda)}\right) e^{i\omega t} \\
&\approx \frac{-8i}{\Omega^2}\left(1 - \frac{2}{i\Omega^2}\frac{(i-1)}{(i+1)}\right) e^{i\omega t} \\
&\approx \frac{8}{\Omega^2}\left(\sin \omega t - i \cos \omega t\right)
\end{aligned} \tag{4.11.13}$$

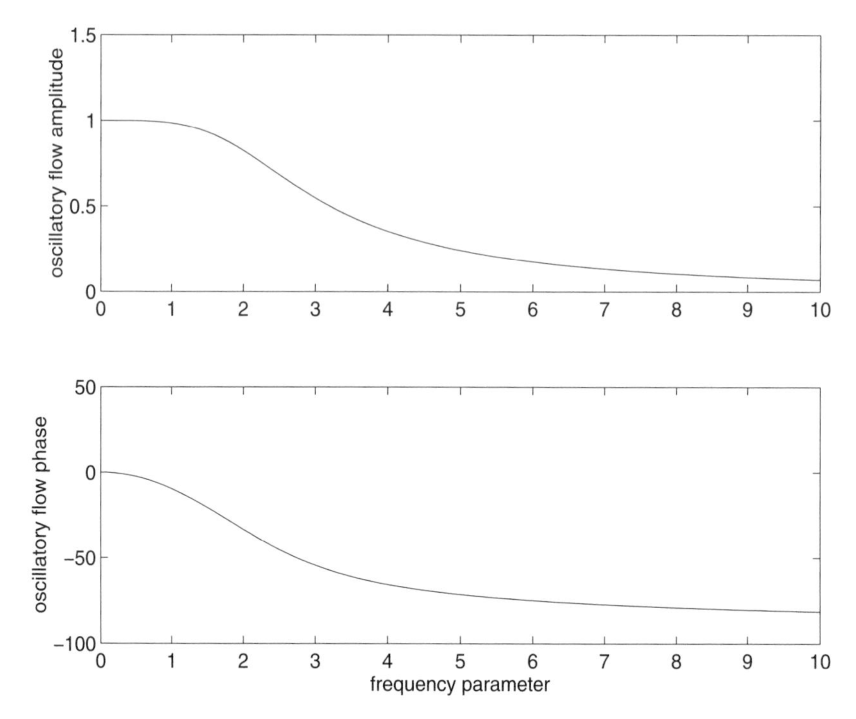

FIGURE 4.11.3. Variation of amplitude and phase of the oscillatory flow rate q_ϕ with frequency parameter Ω. The amplitude (normalized in terms of corresponding steady flow rate) is near 1.0 at low frequency but drops rapidly to near zero as the frequency increases. The phase angle (representing the phase difference between flow rate and pressure gradient, in degrees) is near zero at low frequency but drops rapidly to -90° as the frequency increases.

where the term containing $1/\Omega^2$ is neglected in the last step, and recalling that $\Lambda^2 = -i\Omega^2$. The real and imaginary parts are given by

$$\frac{q_{\phi R}(t)}{q_s} \approx \left(\frac{8}{\Omega^2}\right) \sin \omega t \tag{4.11.14}$$

$$\frac{q_{\phi I}(t)}{q_s} \approx \left(\frac{-8}{\Omega^2}\right) \cos \omega t \tag{4.11.15}$$

These results are similar to those for velocity near the center of the tube (Eq.4.11.8,9), thus flow rate behaves like velocity near the center of the tube, as expected. The results also show how flow rate diminishes at high frequency, as shown in Figs.4.11.2,3.

Corresponding results for shear stress, omitting the details, are given by

$$\frac{\tau_{\phi R}(r,t)}{\tau_s} \approx \left(\frac{\sqrt{2}}{\Omega}\right) (\sin \omega t + \cos \omega t) \tag{4.11.16}$$

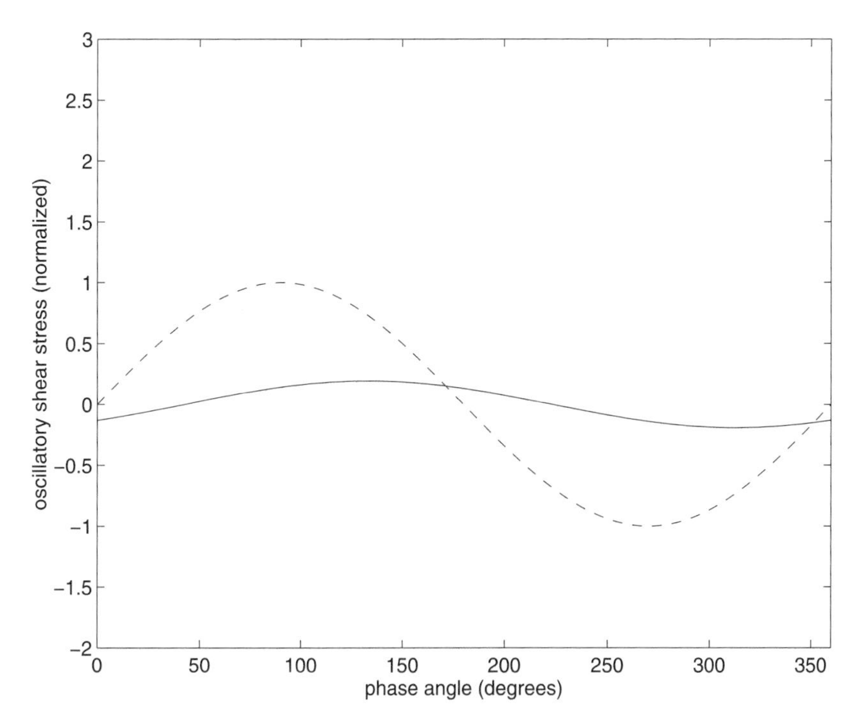

FIGURE 4.11.4. Variation of the imaginary part of oscillatory shear stress $\tau_{\phi I}$ (solid line) compared with the corresponding part of the pressure gradient $k_{\phi I}$ (dashed line), at high frequency ($\Omega = 10$). Shear stress is very low at high frequency, but there is an approximately 90° phase difference between shear stress and pressure gradient.

$$\frac{\tau_{\phi I}(r,t)}{\tau_s} \approx \left(\frac{\sqrt{2}}{\Omega}\right)(\sin\omega t - \cos\omega t) \tag{4.11.17}$$

Variation of shear stress within the oscillatory cycle, and with $\Omega = 10$, is shown in Fig.4.11.4.

For the pumping power, using Eqs.4.9.16,17 and results above for the flow rate, we find

$$\begin{aligned}\frac{H_{\phi I}}{H_s} &= \left(\frac{k_{\phi I}}{k_s}\right)\left(\frac{q_{\phi I}}{q_s}\right) \\ &\approx \frac{-8}{\Omega^2}\sin\omega t\cos\omega t \qquad (4.11.18)\end{aligned}$$

$$\begin{aligned}\frac{H_{\phi R}}{H_s} &= \left(\frac{k_{\phi R}}{k_s}\right)\left(\frac{q_{\phi R}}{q_s}\right) \\ &\approx \frac{8}{\Omega^2}\cos\omega t\sin\omega t \qquad (4.11.19)\end{aligned}$$

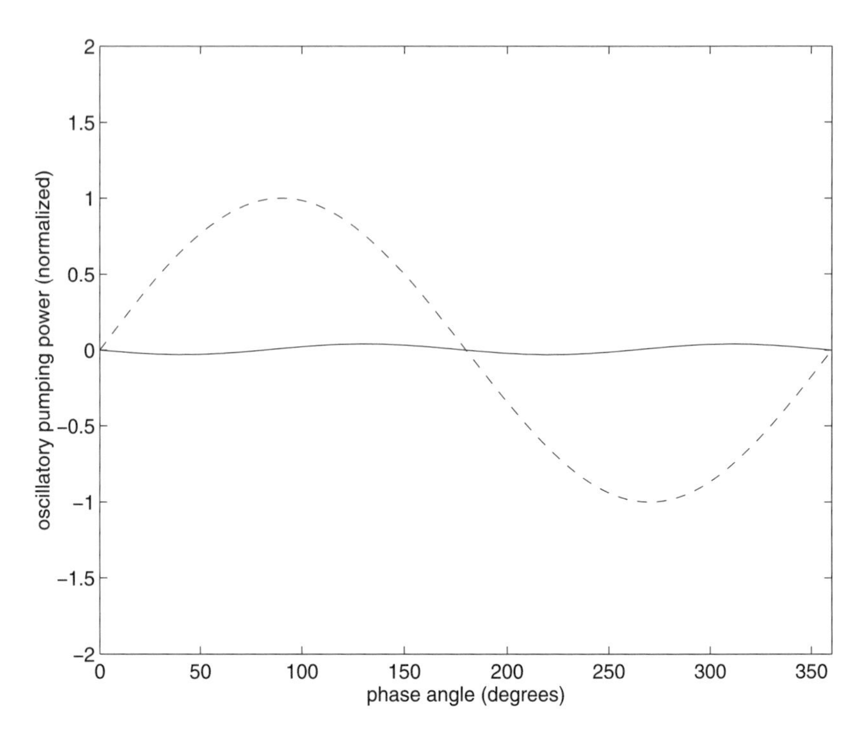

FIGURE 4.11.5. Variation of oscillatory pumping power $H_{\phi I}$ during one cycle (solid line) compared with the corresponding pressure gradient $k_{\phi I}$ (dashed line), at high frequency ($\Omega = 10$). Pumping power is near zero at high frequency, as expected because oscillatory flow rate and shear stress are near zero. By contrast, oscillatory pumping power at *low* frequency is one half the corresponding power in steady flow, even though the *net* flow forward is zero.

In both cases the pumping power vanishes in the limit of very high frequency. Note also, however, that at moderately high frequency where the above expressions can be used for calculating the power, the energy expenditure over one cycle based on these expressions is zero because of Eq.4.10.27. Variation of oscillatory pumping power within the oscillatory cycle, and with $\Omega = 10$, is shown in Fig.4.11.5.

4.12 Oscillatory Flow in Tubes of Elliptic Cross Sections

Pulsatile flow described in this chapter so far relates exclusively to flow in a tube of *circular* cross section. Flow in tubes of noncircular cross sections has not been studied as extensively as that in tubes of circular cross section, and there have been very few studies of *pulsatile* flow in such tubes [11–15]. The problem of pulsatile flow in tubes of *elliptic* cross sections is of particular interest because it offers the possibility of an exact analytical solution. Also, changing the ellipticity of an elliptic cross section produces a wide range of cross sections, including the circular cross section as a special case. Finally, a tube of elliptic cross section provides a good analytical model of a "compressed" blood vessel, which has considerable relevance in pulsatile blood flow.

Solution of the governing equation for pulsatile flow in a tube of elliptic cross section has been obtained in terms of Mathieu functions [11,15,17]. These functions are not as easy to evaluate as Bessel functions, which makes the solution not as readily usable as that for a tube of circular cross section. A brief outline of this solution is presented in this section, with some possible simplifications that make the solution more easy to use, and some results to highlight the effects of ellipticity on pulsatile flow in a tube.

The equation governing the flow is the same as that for a tube of circular cross section, namely, Eq.4.2.7. Boundary conditions are the same as those used for steady flow in a tube of elliptic cross section, namely Eq.3.10.2.

Because the boundary condition is being prescribed on the elliptic boundary of the cross section, solution of Eq.4.2.7 for pulsatile flow seems only possible by transforming to elliptic coordinates [16]

$$\begin{aligned} y &= d\cosh\xi\cos\eta \\ z &= d\sinh\xi\sin\eta \end{aligned} \tag{4.12.1}$$

where y, z are rectangular coordinates in the plane of the elliptic cross section and $2d$ is the distance between the two foci of the ellipse. The curves of constant η represent a family of confocal hyperbolas, while the curves of constant ξ represent a family of confocal ellipses, as illustrated in Fig.4.12.1. The coordinate ξ varies from 0 on the interfocal line to ξ_o on the tube wall. In terms of the elliptic coordinates ξ, η, the governing equation (Eq.4.2.7) with no-slip at the tube wall becomes

$$\frac{2\mu}{d^2(\cosh 2\xi - \cos 2\eta)}\left(\frac{\partial^2 u_{\phi e}}{\partial \xi^2} + \frac{\partial^2 u_{\phi e}}{\partial \eta^2}\right) - \rho\frac{\partial u_{\phi e}}{\partial t} = k_{\phi e}(t) \tag{4.12.2}$$

$$u_{\phi e}(\xi_o, \eta) = 0 \tag{4.12.3}$$

where subscript ϕ is being used as in Section 4.2 to denote oscillatory flow, subscript e is being added as in Section 3.10 to denote elliptic cross section, and $k_{\phi e}$ is the oscillatory pressure gradient in the elliptic tube.

Solution is obtained for an oscillatory pressure gradient as in the case of oscillatory flow in a tube of circular cross section (Eq.4.4.1), that is,

$$k_{\phi e} = k_s e^{i\omega t} \tag{4.12.4}$$

the amplitude k_s being taken the same as that in steady flow in tubes of circular and elliptic cross sections, to facilitate comparison.

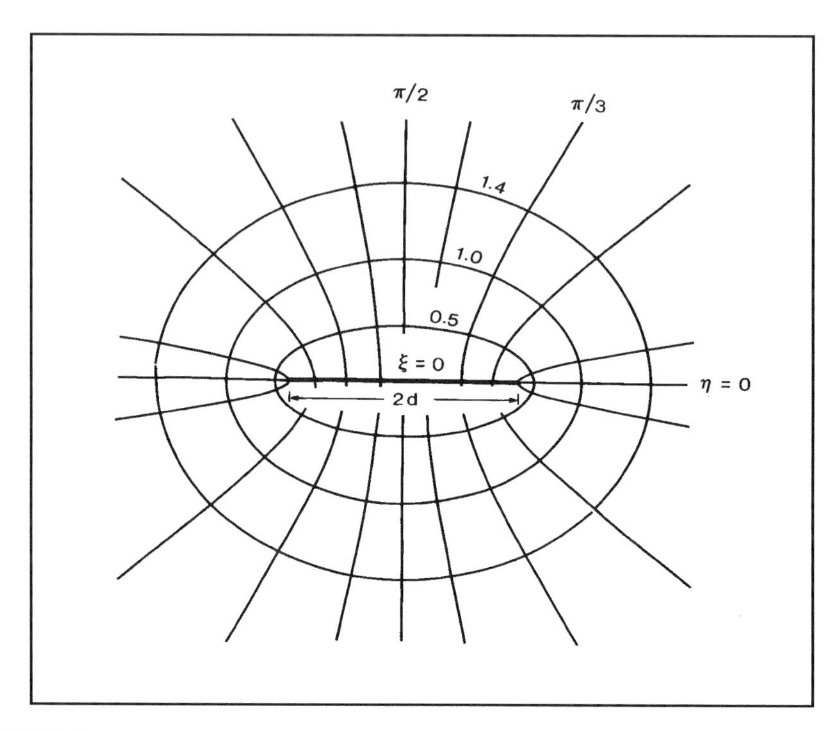

FIGURE 4.12.1. Elliptic coordinate system used in the solution and description of flow in tubes of elliptic cross sections. From [15].

With this choice of pressure gradient the oscillatory part of the velocity takes the form

$$u_{\phi e}(\xi, \eta, t) = U_{\phi e}(\xi, \eta) e^{i\omega t} \tag{4.12.5}$$

and the governing equation finally becomes an equation for $U_{\phi e}$, namely,

$$\frac{2}{d^2(\cosh 2\xi - \cos 2\eta)} \left(\frac{\partial^2 U_{\phi e}}{\partial \xi^2} + \frac{\partial^2 U_{\phi e}}{\partial \eta^2} \right) - \frac{i\rho\omega}{\mu} U_{\phi e} = \frac{k_s}{\mu} \tag{4.12.6}$$

A nondimensional frequency parameter is defined by

$$\Omega_e = \sqrt{\frac{\rho\omega}{\mu}} \sigma \tag{4.12.7}$$

where

$$\sigma = \sqrt{\frac{2b^2c^2}{b^2 + c^2}} \tag{4.12.8}$$

and b, c are semi-minor and semi-major axes of the ellipse as in the steady-flow case.

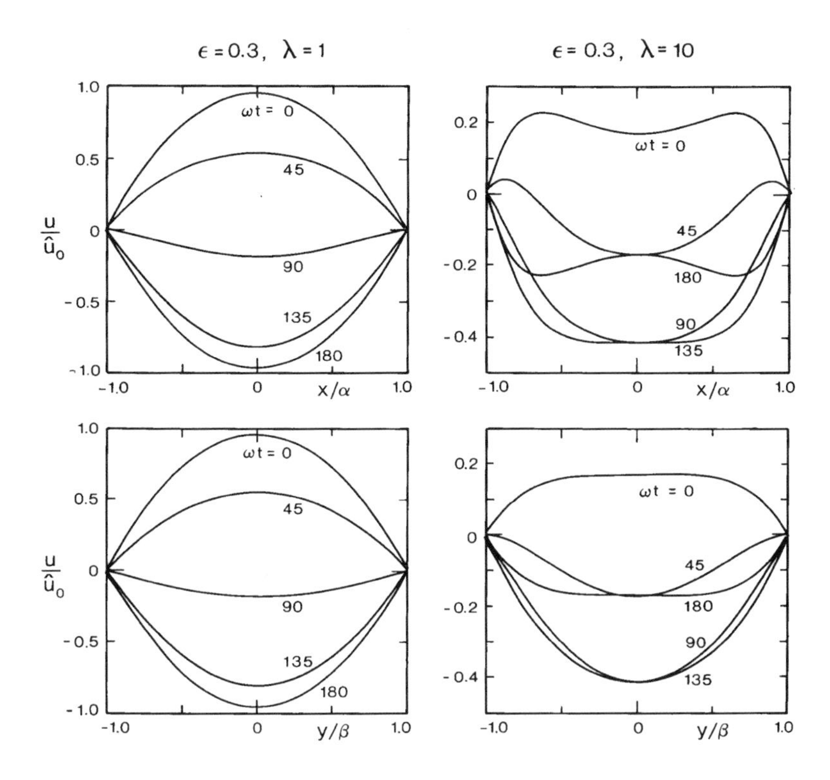

FIGURE 4.12.2. Oscillatory velocity profiles along the major axis (top) and minor axis (bottom) of an elliptic cross section at low ($\lambda = 1$, left) and moderately high frequency ($\lambda = 10$, right). In each panel, velocity profiles are shown at different phase angles ωt within the oscillatory cycle, ranging from $\omega t = 0°$ to $\omega t = 180°$, the second half of the cycle being omitted because of symmetry. The coordinates x, y and the axes α, β in this figure, from [15], correspond to coordinates y, z and axes b, c used in this book. The parameter λ corresponds to the square of the frequency parameter Ω_e used in this book. The effects of ellipticity on the flow are here seen in terms of the difference between velocity profiles along the major axis in the top panels and minor axis in the bottom panels. The difference is seen to be insignificant at low frequency, but becomes considerable at high frequency.

Equation 4.12.6 has a solution of the form [15]

$$U_{\phi e}(\xi, \eta) = \frac{4\hat{u}_{se}}{i\Omega^2} + \sum_{n=0}^{\infty} C_{2n} Ce_{2n}(\xi, -\gamma)\, ce_{2n}(\eta, -\gamma) \tag{4.12.9}$$

where C_{2n} is a constant determined by the boundary condition, ce_{2n} and Ce_{2n} are the ordinary and modified Mathieu functions [17], and

$$\gamma = \frac{i\rho\omega d^2}{4\mu} \tag{4.12.10}$$

Some velocity profiles are shown in Fig.4.12.2 and oscillatory flow rate in Fig.4.12.3.

Evaluation of this solution is highly cumbersome because of the infinite series in Eq.4.12.9 and because of the difficulties involved in the evaluation of Mathieu functions in general. Some simplifications are possible, however, under certain conditions.

At *low frequency* it is found that velocity and flow rate can be put in the form

$$u_{\phi e}(y, z, t) \approx u_{se}(y, z)e^{i\omega t} \tag{4.12.11}$$

$$q_{\phi e}(t) \approx q_{se}e^{i\omega t} \tag{4.12.12}$$

where u_{se} and q_{se} are the corresponding velocity and flow rate in *steady* flow in a tube of elliptic cross section (Eqs.3.10.3,4).

At *low ellipticity* ($\epsilon = b/c$) it is found that velocity and flow rate become very close to those in a tube of circular cross section, with the radius of the tube being replaced by σ. In fact, for $\epsilon > 0.9$ it is found that differences from the circular case are negligibly small.

Finally, it has been found that the ratio $q_{\phi e}/q_{se}$ at *all frequencies* is highly insensitive to the value of ellipticity ϵ. In particular, therefore, this ratio for a tube of elliptic cross section is approximately equal to the corresponding ratio for a tube of circular cross section, that is,

$$\frac{q_{\phi e}(t)}{q_{se}} \approx \frac{q_\phi(t)}{q_s} \tag{4.12.13}$$

the flow rates on the right-hand side being those for a tube of circular cross section (Eqs.3.4.3, 4.7.5). This permits the following approximation for the flow rate in a tube of elliptic cross section, using Eq.4.7.5

$$\frac{q_{\phi e}(t)}{q_{se}} \approx \frac{-8}{\Lambda_e^2}\left(1 - \frac{2J_1(\Lambda_e)}{\Lambda_e J_0(\Lambda_e)}\right)e^{i\omega_e t} \tag{4.12.14}$$

where

$$\Lambda_e = \left(\frac{i-1}{\sqrt{2}}\right)\Omega_e \tag{4.12.15}$$

and Ω_e, which contains the parameter σ replacing the radius of the circular cross section, is defined by Eq.4.12.7. This approximate expression for the flow rate is clearly much easier to use than one derived from Eq.4.12.9, and the approximation is equally valid for the full range of frequency and ellipticity.

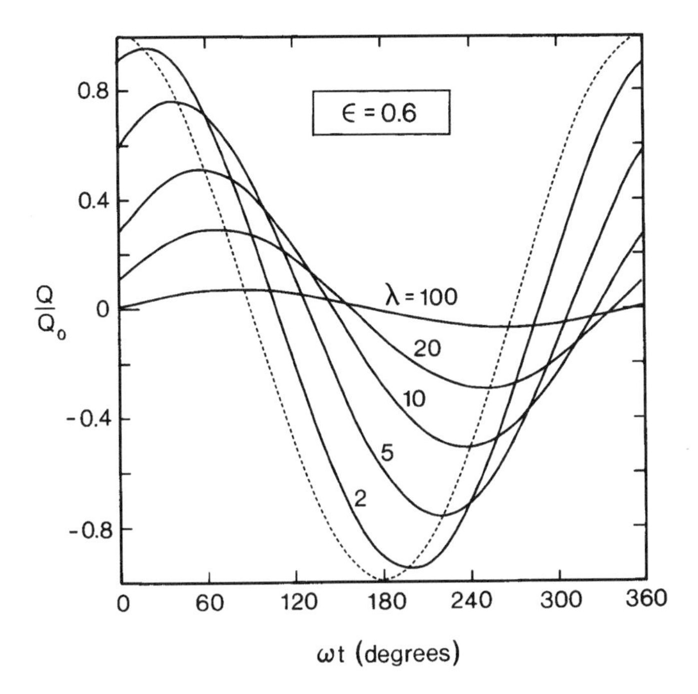

FIGURE 4.12.3. Oscillatory flow rate in a tube of elliptic cross section of ellipticity $\epsilon = 0.6$, normalized in terms of the corresponding steady-flow rate. The ratio Q/Q_0 and parameter λ in this figure, from [15], correspond to the ratio $q_{\phi e}/q_{se}$ and the square of the frequency parameter Ω_e used in this book. Each curve represents variation of the flow rate within one oscillatory cycle for a given value of the frequency parameter. The dotted curve represents the corresponding pressure gradient. As in the case of a tube of circular cross section, flow rate is in phase with pressure gradient at low frequency, but becomes increasingly out of phase and diminishes in magnitude as the frequency increases.

4.13 Problems

1. Explain a principal feature of oscillatory flow in a rigid tube that makes it somewhat artificial physiologically.
2. State the main assumption that makes it possible to separate the equations governing the steady and oscillatory parts of pulsatile flow in a rigid tube, as in Eq.4.2.2, and explain the restrictions that this places on the flow being considered.
3. Instead of two separate sine and cosine series, the harmonics of a composite oscillatory wave can always be put in the form of only a cosine series in which each term has an amplitude C and a phase angle Φ, that

is, in the notation of Eq.4.3.2

$$f(t) = \sum_{n=1}^{\infty} C_n \cos\left(\frac{2n\pi t}{T} + \Phi_n\right)$$

The first 10 harmonics of the composite wave shown in Fig.4.3.1 have the following amplitudes and phase angles (in degrees)

7.5803	-173.9168
5.4124	88.9222
1.5210	-21.7046
0.5217	-33.5372
0.8311	-126.8094
0.6851	135.0559
0.2584	152.0862
0.5408	44.0552
0.2715	-72.0738
0.0991	11.3354

Write down the expressions required to compute the first four harmonics shown in Fig.4.3.2, and indicate the way in which approximations for the composite wave shown in Fig.4.3.1 would be constructed from these.

4. Show by substitution that separation of variables in Eq.4.4.3 reduces the *partial* differential equation (Eq.4.4.2) into the *ordinary* differential equation (Eq.4.4.4).
5. The first term on the right-hand side in Eq.4.5.1 for the solution of Bessel equation represents a *particular* solution of Eq.4.4.4. Show by substitution that

$$U_\phi(r) = \frac{ik_s a^2}{\mu\Omega^2}$$

does satisfy that equation.

6. The second and third terms on the right-hand side in Eq.4.5.1 for the solution of Bessel equation represent two independent solutions of the *homogeneous* form of Eq.4.4.4. Show by substitution that each of

$$U_\phi(r) = AJ_0(\zeta), \;\; U_\phi(r) = BY_0(\zeta)$$

satisfies the homogeneous form of Eq.4.4.4.

7. Consider pulsatile flow in a rigid tube with frequency $\Omega = 3.0$ and driving pressure gradient of the form $k_s \cos\omega t$. Using Eq.4.6.2 and values of Bessel functions in Appendix A, find the velocity at the center of the tube (a) at the beginning of the cycle and (b) at a quarter way through the cycle. Compare your results with those in Fig.4.6.1.
8. Consider pulsatile flow in a rigid tube with frequency $\Omega = 3.0$ and driving pressure gradient of the form $k_s \sin\omega t$. Using Eq.4.7.7 and values of Bessel functions in Appendix A, find the peak flow rate, which, accord-

ing to Fig.4.7.1, occurs at $\omega t \approx 150°$. Compare your result with that in Fig.4.7.1.

9. Results in Fig.4.9.1 indicate that in oscillatory flow the average rate of energy dissipation at the tube wall is nonzero, therefore requiring pumping power to maintain. In *pulsatile* flow, consisting of oscillatory plus steady flow components, this "wasted" pumping power would be required in addition to that required for the steady part of the flow. Use Eqs.4.9.16,17 to calculate the average of this wasted power over one oscillatory cycle, expressed as a fraction of the corresponding pumping power in steady flow. Use $\Omega = 3.0$ in the calculation, then compare the result with that in Fig.4.9.1

10. Use approximate expressions for low and for high frequency to compare the peak flow rates reached at $\Omega = 1$ and $\Omega = 10$, and compare the results visually with those in Fig.4.11.3.

11. From the results of the previous example, deduce the phase lag of flow behind pressure gradient at low ($\Omega = 1$) and high ($\Omega = 10$) frequency. Compare the results visually with those in Fig.4.11.3.

12. The back and forth movements of the fluid in oscillatory flow waste energy not because of acceleration and deceleration of the fluid but because of viscous dissipation at the tube wall. Use the results of Sections 4.10 and 4.11 to determine if this wasteful energy expenditure is higher at high or low frequency, and determine its magnitude at $\Omega = 1$ and $\Omega = 10$, expressed as a fraction of the corresponding pumping power in steady flow.

4.14 References and Further Reading

1. Lighthill M, 1975. Mathematical Biofluiddynamics. Society for Industrial and Applied Mathematics, Philadelphia.
2. Walker JS, 1988. Fourier Analysis. Oxford University Press, New York.
3. Brigham EO, 1988. The Fast Fourier Transform and its Applications. Prentice Hall, Englewood Cliffs, New Jersey.
4. McLachlan NW, 1955. Bessel Functions for Engineers. Clarendon Press, Oxford.
5. Watson GN, 1958. A Treatise on the Theory of Bessel Functions. Cambridge University Press. Cambridge.
6. Sexl T, 1930. Über den von E.G. Richardson entdeckten "Annulareffekt." Zeitschrift für Physik 61:349–362.
7. Womersley JR, 1955. Oscillatory motion of a viscous liquid in a thin-walled elastic tube—I: The linear approximation for long waves. Philosophical Magazine 46:199–221.
8. Uchida S, 1956. The pulsating viscous flow superimposed on the steady laminar motion of incompressible fluid in a circular pipe. Zeitschrift für angewandte Mathematik und Physik 7:403-422.

9. McDonald DA, 1974. Blood flow in arteries. Edward Arnold, London.
10. Milnor WR, 1989. Hemodynamics. Williams and Wilkins, Baltimore.
11. Khamrui SR, 1957. On the flow of a viscous liquid through a tube of elliptic section under the influence of a periodic gradient. Bulletin of the Calcutta Mathematical Society 49:57–60.
12. Begum R, Zamir M, 1990. Flow in tubes of non-circular cross sections. In: Rahman M (ed), Ocean Waves Mechanics: Computational Fluid Dynamics and Mathematical Modeling. Computational Mechanics Publications, Southampton.
13. Duan B, Zamir M, 1991. Approximate solution for pulsatile flow in tubes of slightly noncircular cross-sections. Utilitas Mathematica 40:13–26.
14. Quadir R, Zamir M, 1997. Entry length and flow development in tubes of rectangular and elliptic cross sections. In: Rahman M (ed), Laminar and Turbulent Boundary Layers. Computational Mechanics Publications, Southampton.
15. Haslam M, Zamir M, 1998. Pulsatile flow in tubes of elliptic cross sections. Annals of Biomedical Engineering 26:1–8.
16. Moon PH, Spencer DE, 1961. Field Theory for Engineers. Van Nostrand, Princeton, New Jersey.
17. McLachlan NW, 1964. Theory and Application of Mathieu Functions. Dover Publications, New York.

5

Pulsatile Flow in an Elastic Tube

5.1 Introduction

In the case of a *rigid* tube it is possible to postulate a fully developed region away from the tube entrance where the flow is independent of x, thus derivatives of u, v with respect to x are zero. The equation of continuity combined with the boundary condition $v=0$ at the tube wall then leads to v being identically zero and the governing equations reduce to (Eq.3.2.9)

$$\rho \frac{\partial u}{\partial t} + \frac{\partial p}{\partial x} = \mu \left(\frac{\partial^2 u}{\partial r^2} + \frac{1}{r} \frac{\partial u}{\partial r} \right)$$

As we saw in Chapter 4, the pressure gradient term in that case is a function of t only, not of x, and the velocity u is then a function of r, t only, not of x. For an oscillatory pressure gradient the velocity u then oscillates with the same frequency, and since it is not a function of x, it will represent the velocity at every cross section of the tube. The fluid then oscillates *in bulk*, or "en mass." *There is no wave motion* when the tube is rigid.

When the tube is *nonrigid*, because of movement of the tube wall the radial velocity v of the fluid can no longer be identically zero, and both u and v can no longer be independent of x even away from the tube entrance (Fig.5.1.1). Equation 3.2.9 is then no longer valid and we must return to the full Navier–Stokes equations, assuming only axial symmetry, that is (Eqs.3.2.2–4)

$$\rho \left(\frac{\partial u}{\partial t} + u \frac{\partial u}{\partial x} + v \frac{\partial u}{\partial r} \right) \quad + \quad \frac{\partial p}{\partial x}$$

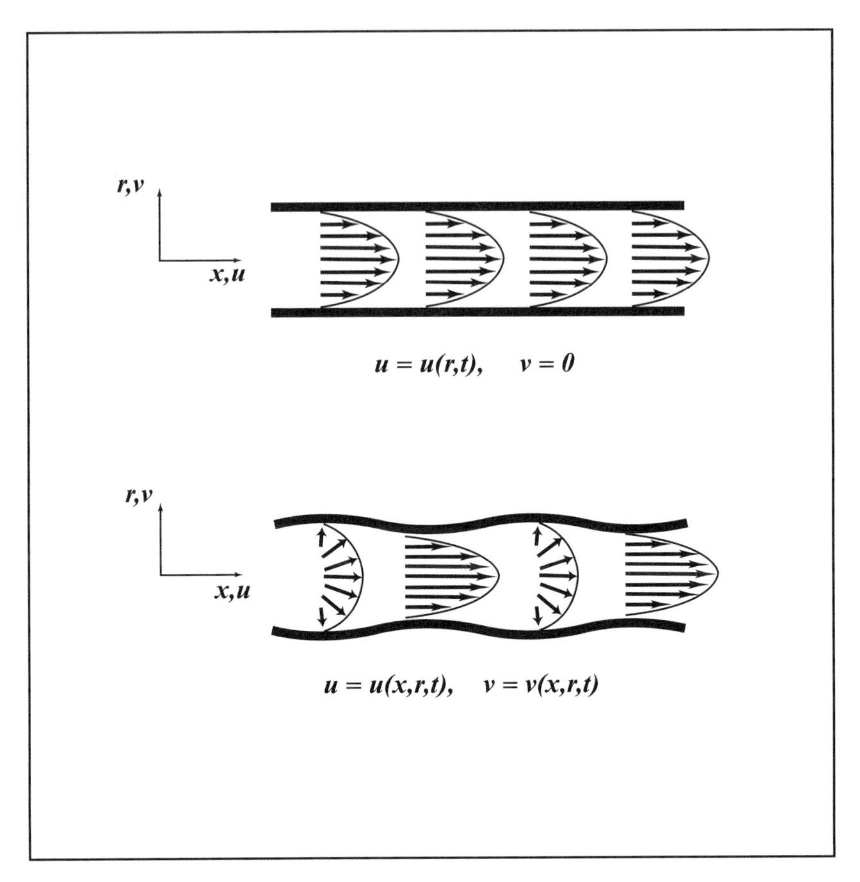

FIGURE 5.1.1. Wave propagation in an elastic tube. In the fully developed region of flow in a *rigid* tube (top) oscillatory pressure changes occur simultaneously at every point along the tube to the effect that the fluid oscillates in bulk. There is no wave motion. In an elastic tube (bottom) pressure changes cause *local* movements of the fluid and tube wall, which then propagate downstream in the form of a wave. Velocity is no longer independent of x, and the radial velocity v is no longer zero.

$$= \mu \left(\frac{\partial^2 u}{\partial x^2} + \frac{\partial^2 u}{\partial r^2} + \frac{1}{r}\frac{\partial u}{\partial r} \right)$$

$$\rho \left(\frac{\partial v}{\partial t} + u\frac{\partial v}{\partial x} + v\frac{\partial v}{\partial r} \right) + \frac{\partial p}{\partial r}$$
$$= \mu \left(\frac{\partial^2 v}{\partial x^2} + \frac{\partial^2 v}{\partial r^2} + \frac{1}{r}\frac{\partial v}{\partial r} - \frac{v}{r^2} \right)$$

$$\frac{\partial u}{\partial x} + \frac{\partial v}{\partial r} + \frac{v}{r} = 0$$

The most important difference between these equations and that for a rigid tube is that here u and v are functions of x. The input pressure gradient at the tube entrance is a function of t, but within the tube the pressure gradient and two velocities are functions of both x and t. An input oscillatory pressure at the tube entrance now *propagates* down the tube. *There is wave motion within the tube.*

The speed with which the wave propagates down the tube depends on the elasticity of the tube wall. If the wall thickness is small compared with the tube radius, and if the effects of viscosity can be neglected, the wave speed is given approximately by the so called Moen–Korteweg formula [1–4]

$$c_0 = \sqrt{\frac{Eh}{\rho d}} \tag{5.1.1}$$

where E is Young's modulus for the tube wall and h is its thickness, d is the diameter of the tube and ρ is constant density of the fluid. The latter is important because if ρ is not constant then changes in pressure lead to compression and expansion of the *fluid* within the tube, which provides another mechanism for wave propagation that can occur even in a rigid tube.

Higher values of E represent lower elasticity of the tube wall, E being infinite when the wall is rigid. Thus the bulk motion of fluid in the case of oscillatory flow in a rigid tube can be thought of as resulting from a wave traveling with infinite speed so that a pressure change at one end of the tube reaches the other end instantaneously. In other words, as the driving pressure oscillates in time, pressure changes occur simultaneously at every point along the tube. In the case of an *elastic* tube, by contrast, pressure changes take effect locally at first and then propagate downstream [5,6].

Some simplification of the governing equations is possible if it can be assumed that the length L of the propagating wave is much higher than the tube radius a (long-wave approximation), and wave speed c_0 is much higher than the average flow velocity $\overline{u}$ within the tube, that is, if

$$\frac{a}{L}, \frac{\overline{u}}{c_0} << 1 \tag{5.1.2}$$

Under these conditions the following comparison of terms in Eqs.3.2.2-4 applies

$$u\frac{\partial u}{\partial x}, v\frac{\partial u}{\partial r} << \frac{\partial u}{\partial t}$$

$$u\frac{\partial v}{\partial x}, v\frac{\partial v}{\partial r} << \frac{\partial v}{\partial t}$$

$$\frac{\partial^2 u}{\partial x^2} << \frac{\partial^2 u}{\partial r^2}$$

$$\frac{\partial^2 v}{\partial x^2} << \frac{\partial^2 v}{\partial r^2} \tag{5.1.3}$$

and the governing equations thus reduce to

$$\rho \frac{\partial u}{\partial t} + \frac{\partial p}{\partial x} = \mu \left(\frac{\partial^2 u}{\partial r^2} + \frac{1}{r} \frac{\partial u}{\partial r} \right) \tag{5.1.4}$$

$$\rho \frac{\partial v}{\partial t} + \frac{\partial p}{\partial r} = \mu \left(\frac{\partial^2 v}{\partial r^2} + \frac{1}{r} \frac{\partial v}{\partial r} - \frac{v}{r^2} \right) \tag{5.1.5}$$

$$\frac{\partial u}{\partial x} + \frac{\partial v}{\partial r} + \frac{v}{r} = 0 \tag{5.1.6}$$

There are three equations for three dependent variables, namely, $u(x, r, t)$, $v(x, r, t)$, $p(x, r, t)$, compared with only one equation for $u(r, t)$ in the case of a rigid tube.

5.2 Bessel Equations and Solutions

As in the case of a rigid tube, a solution of the governing equations is possible when the input pressure at the tube entrance is a simple "sinusoidal" oscillatory function. In this case, however, as discussed previously, the resulting pressure and flow distributions within the tube are oscillatory both in space and time. At any point in time, the pressure and flow distributions are sinusoidal in x, and at any fixed position they are sinusoidal in t. As in the case of a rigid tube, the analysis is considerably easier if the oscillations are considered as complex exponentials rather than sine or cosine functions. Mathematically, then, the simplified governing equations are found to have a solution for which the three dependent variables are of the form

$$p(x, r, t) = P(r)e^{i\omega(t-x/c)} \tag{5.2.1}$$

$$u(x, r, t) = U(r)e^{i\omega(t-x/c)} \tag{5.2.2}$$

$$v(x, r, t) = V(r)e^{i\omega(t-x/c)} \tag{5.2.3}$$

As in the case of a rigid tube, ω is the frequency of oscillation of the input pressure, and as in that case, the oscillations *within* the tube have the same frequency. The analytical advantage of the complex exponential form is noted upon substitution of these into Eqs.5.1.4–6 with the result that the exponential terms cancel throughout, leaving *ordinary* differential equations for $P(r), U(r), V(r)$, namely,

$$\frac{d^2 U}{dr^2} + \frac{1}{r} \frac{dU}{dr} - \frac{i\rho\omega}{\mu} U = -\frac{i\omega}{\mu c} P \tag{5.2.4}$$

$$\frac{d^2V}{dr^2} + \frac{1}{r}\frac{dV}{dr} - \left(\frac{1}{r^2} + \frac{i\rho\omega}{\mu}\right)V = \frac{1}{\mu}\frac{dP}{dr} \tag{5.2.5}$$

$$\frac{dV}{dr} + \frac{V}{r} - \frac{i\omega}{c}U = 0 \tag{5.2.6}$$

The first two of these equations are forms of Bessel equations with known solutions in terms of Bessel functions [7,8]. To put the equations in standard form we introduce, as in the case of a rigid tube (Eqs.4.4.5, 4.5.4,5),

$$\Omega = \sqrt{\frac{\rho\omega}{\mu}}\,a, \quad \Lambda = \left(\frac{i-1}{\sqrt{2}}\right)\Omega, \quad \zeta = \Lambda\frac{r}{a}$$

In terms of ζ, the three governing equations now become

$$\frac{d^2U}{d\zeta^2} + \frac{1}{\zeta}\frac{dU}{d\zeta} + U = \frac{1}{\rho c}P \tag{5.2.7}$$

$$\frac{d^2V}{d\zeta^2} + \frac{1}{\zeta}\frac{dV}{d\zeta} + \left(1 - \frac{1}{\zeta^2}\right)V = \frac{i\Lambda}{\rho a\omega}\frac{dP}{d\zeta} \tag{5.2.8}$$

$$\frac{dV}{d\zeta} + \frac{V}{\zeta} - \frac{i\omega a}{c\Lambda}U = 0 \tag{5.2.9}$$

The required boundary conditions are zero velocities at the tube wall and finite velocity at the tube center. Because the tube wall is in motion, the first of these boundary conditions presents a severe difficulty, which makes it impossible to obtain an analytical solution of the equations. As a reasonable approximation the boundary condition is applied at a fixed radius a, which is taken to be the *neutral position of the tube wall.* Thus the approximate boundary conditions become

$$\begin{aligned} r = a,\ \ \zeta = \Lambda: &\quad U(a), V(a) = 0 \\ r = 0,\ \ \zeta = 0: &\quad U(0), V(0)\ \text{finite} \end{aligned} \tag{5.2.10}$$

Solutions of the first two governing equations that satisfy the third equation as well as these boundary conditions are given by

$$U(r) = AJ_0(\zeta) + B\frac{a\gamma}{\mu(i\Omega^2 + \gamma^2)}J_0\left(\frac{\gamma}{\Lambda}\zeta\right) \tag{5.2.11}$$

$$V(r) = A\frac{\gamma}{\Lambda}J_1(\zeta) + B\frac{a\gamma}{\mu(i\Omega^2 + \gamma^2)}J_1\left(\frac{\gamma}{\Lambda}\zeta\right) \tag{5.2.12}$$

$$P(r) = BJ_0\left(\frac{\gamma r}{a}\right) \tag{5.2.13}$$

where A, B are arbitrary constants and

$$\gamma = \frac{i\omega a}{c} \tag{5.2.14}$$

The solutions can be simplified by noting that

$$\gamma \sim \left(\frac{a}{L}\right) << 1 \tag{5.2.15}$$

$$J_0\left(\frac{\gamma r}{a}\right) \approx 1 \tag{5.2.16}$$

$$J_1\left(\frac{\gamma r}{a}\right) \approx \frac{1}{2}\frac{\gamma r}{a} \tag{5.2.17}$$

Applying these approximations the solutions reduce to

$$U(r) = AJ_0(\zeta) + B\left(\frac{1}{\rho c}\right) \tag{5.2.18}$$

$$V(r) = A\frac{i\omega a}{c\Lambda}J_1(\zeta) + B\left(\frac{i\omega r}{2\rho c^2}\right) \tag{5.2.19}$$

$$P(r) = B \ \ \text{(constant)} \tag{5.2.20}$$

The constants A, B are determined by matching fluid and wall velocities at $r = a$, which requires that now the motion of the tube wall be considered.

5.3 Balance of Forces

Elastic movements of the tube wall are governed by the equations of elasticity which in their most general form are considerably more complicated than the equations of fluid flow. The reason for this is that in fluid flow one is usually concerned with three velocity components and pressure, while in the case of elasticity one may be concerned with three displacement components and six internal stresses. To deal with the elasticity problem in its most general form is far more than is required for the present purpose. In what follows, therefore, rather than derive the equations of elasticity in their general form, we simply consider the forces acting on an element of the tube wall, then extract from the theory of elasticity only what is required for the purpose at hand.

Consider an element of the tube wall of thickness h $(=\delta r)$, arc length $a\delta\theta$, where a is the neutral radius of the tube, and axial length δx (Fig.5.3.1). The volume and mass of the element are then respectively given by

$$\delta V \approx ha\delta\theta\delta x \tag{5.3.1}$$

$$\delta m \approx \rho_w \delta V \tag{5.3.2}$$

where ρ_w is density of the tube wall.

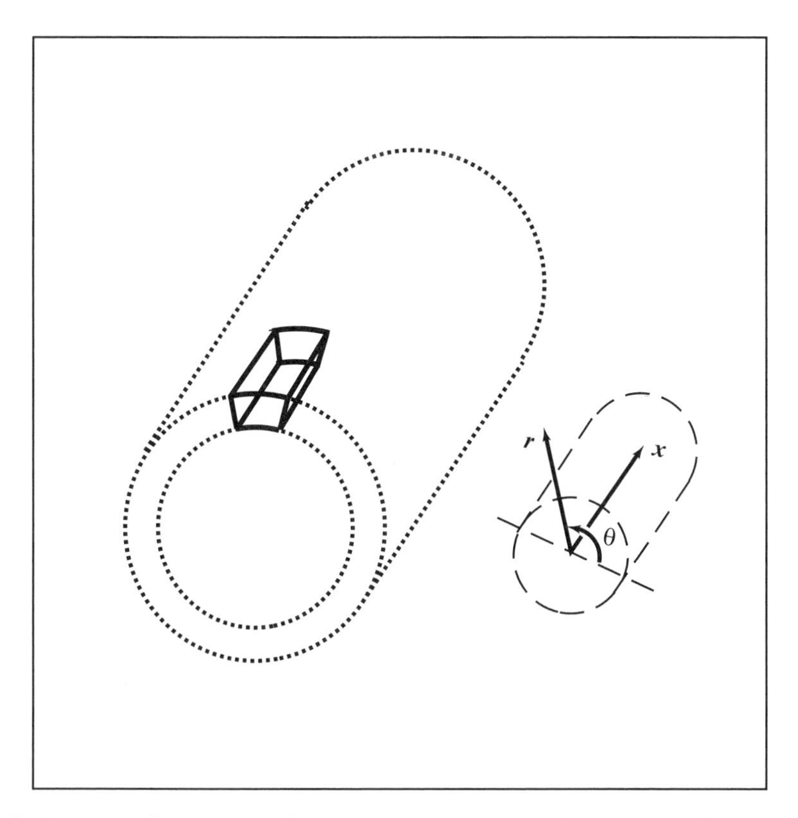

FIGURE 5.3.1. An element of the tube wall considered for analysis of movement of the tube wall. Dimensions of the element in the three coordinate directions are δx, δr, and $a\delta\theta$, where a is the tube radius. On the assumption that the wall thickness h is small compared with a, we take $\delta r = h$, which implies that radial gradients within the tube wall are neglected.

Forces acting on this element of the tube wall result from four mechanical stresses, each having the dimensions of *force per unit area* (Fig.5.3.2).

(i) Axial tension within the tube wall, to be denoted by S_{xx}. This tension is in general a function of x, thus leading to a force in the positive x direction due to a change δS_{xx} over the length of the element, given by

$$\delta S_{xx} \times ha\delta\theta = \frac{\partial S_{xx}}{\partial x}\delta x \times ha\delta\theta \tag{5.3.3}$$

(ii) Radial stress, to be denoted by S_{rr}, resulting from *angular* tension within the tube wall, and producing a force pushing the tube wall toward the center of the tube, given by

$$-S_{rr} \times a\delta\theta\delta x \tag{5.3.4}$$

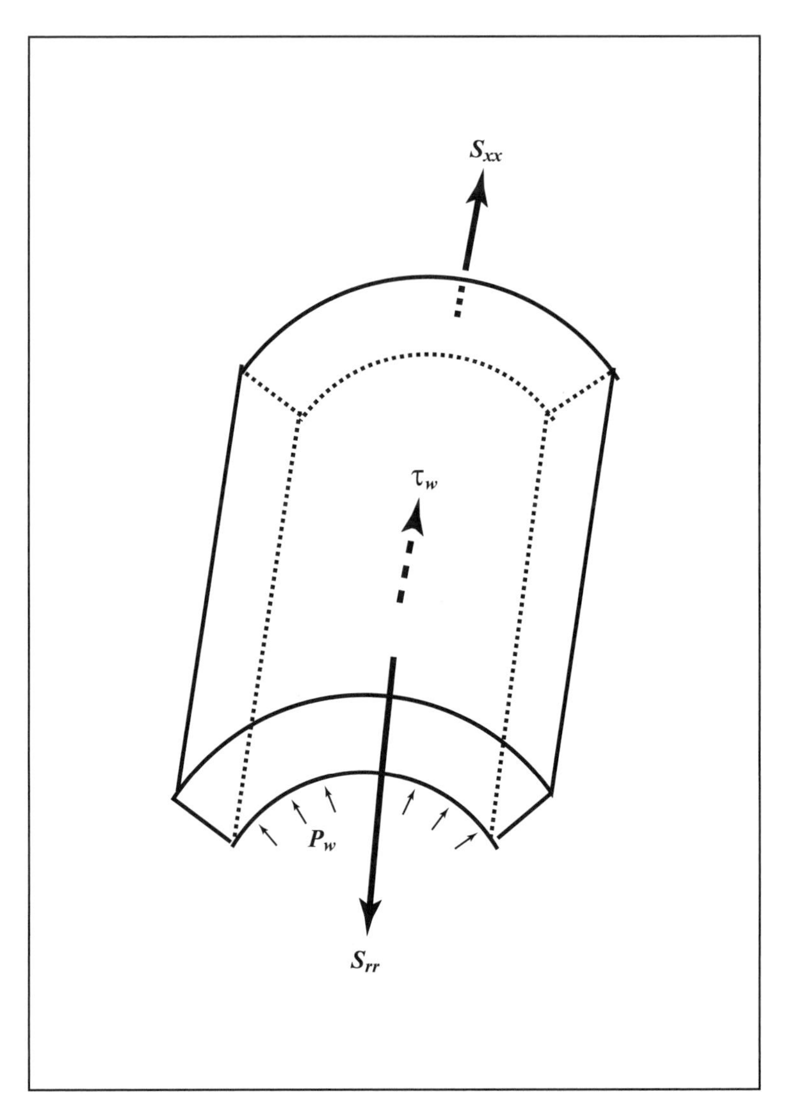

FIGURE 5.3.2. Equations governing motion of the tube wall are based on a balance of forces acting on an element of the tube wall (Fig.5.3.1), shown here enlarged. The forces arise from four stresses (which have the dimensions of *force/area*): axial tension S_{xx} within the tube wall, radial tension S_{rr} pulling elements of the tube wall toward the tube axis and arising from angular tension $S_{\theta\theta}$ in the tube wall (Fig.5.3.3), shear stress τ_w exerted by the fluid on the inner surface of the tube, and pressure p_w exerted radially by the fluid on the inner surface of the tube.

While S_{rr} may in general vary within the thickness of the tube wall, that is it may be a function of r, thus producing another part of the radial force due to a change δS_{rr} over the thickness of the tube wall, this part is being neglected here on the assumption that the tube wall is thin.

(iii) Fluid pressure within the tube, being the net difference between pressures acting on the inside and outside of the tube wall, to be denoted by p_w, and leading to a force in the positive r direction given by

$$p_w \times a\delta\theta\delta x \tag{5.3.5}$$

(iv) Shear stress τ_w exerted by the moving fluid on the tube wall and leading to a force in the flow direction given by

$$\tau_w \times a\delta\theta\delta x \tag{5.3.6}$$

The net force in each of the three coordinate directions must equal the acceleration of the element in that direction times its mass, thus providing an equation of motion in each direction. If ξ, η, ϕ represent displacements of the tube wall element in the x, r, θ directions, respectively, then in the axial direction we have

$$\rho_w \times ha\delta\theta\delta x \times \frac{d^2\xi}{dt^2} = ha\delta\theta \times \frac{\partial S_{xx}}{\partial x}\delta x + a\delta\theta\delta x \times \tau_w \tag{5.3.7}$$

which simplifies to

$$\rho_w h \frac{d^2\xi}{dt^2} = h\frac{\partial S_{xx}}{\partial x} + \tau_w \tag{5.3.8}$$

Similarly, in the radial direction we have

$$\rho_w \times ha\delta\theta\delta x \times \frac{d^2\eta}{dt^2} = a\delta\theta\delta x \times p_w - a\delta\theta\delta x \times S_{rr} \tag{5.3.9}$$

which simplifies to

$$\rho_w h \frac{d^2\eta}{dt^2} = p_w - S_{rr} \tag{5.3.10}$$

In the angular direction acceleration is zero because of axial symmetry and because of the absence of any external force in that direction. As stated earlier, however, because of curvature of the tube wall the internal angular stress $S_{\theta\theta}$ produces not only radial strain (change in wall thickness) but also a *movement* of the wall in the radial direction. The latter is caused by change in the tube radius, which in turn is caused by change in the circumference of the tube circular cross section, that is by angular strain, $S_{\theta\theta}$. In fact a useful relation between the angular and radial strains can be

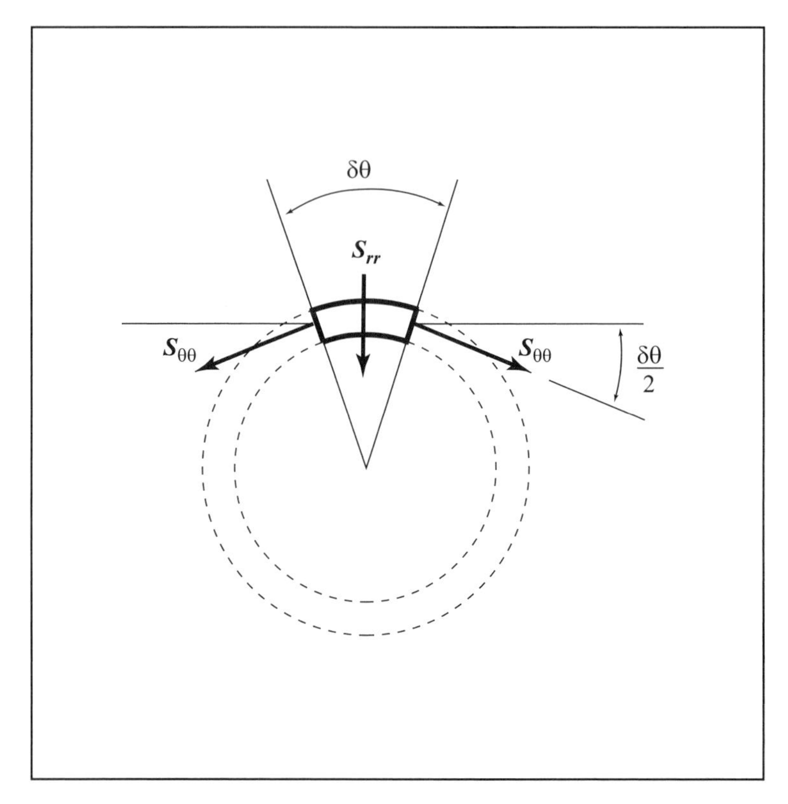

FIGURE 5.3.3. Radial stress S_{rr}, which acts to pull the tube wall toward the axis of the tube, is related to angular tension $S_{\theta\theta}$ within the tube wall (Eq.5.3.12).

obtained by equating forces in the radial direction for a small segment of the wall when in a state of equilibrium (Fig.5.3.3), namely,

$$\begin{aligned} a\delta\theta \times S_{rr} &= 2 \times h \times S_{\theta\theta} \times \sin\left(\frac{\delta\theta}{2}\right) \\ &\approx hS_{\theta\theta}\delta\theta \end{aligned} \tag{5.3.11}$$

which gives

$$S_{rr} = \frac{h}{a} S_{\theta\theta} \tag{5.3.12}$$

5.4 Equations of Wall Motion

Stresses $(S_{xx}, S_{rr}, S_{\theta\theta})$ in the equations of wall motion (Eqs.5.3.8,10) must be expressed in terms of the displacements (ξ, η) in order to solve these equations. This is achieved by stress–strain relations that are found to

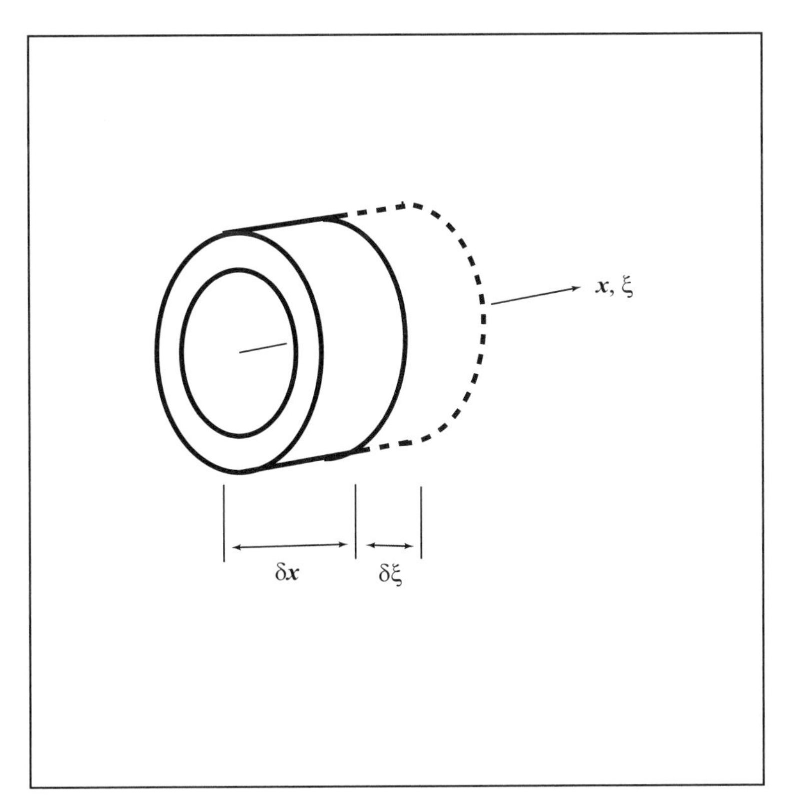

FIGURE 5.4.1. Axial displacement ξ is in general different at different points along the tube wall. As a result, an element of original length δx may stretch by an amount $\delta\xi$, where $\delta\xi$ is the change in ξ over the length of the element.

exist in an elastic body [9–11]. The relations are analogous to the relations between stresses and *rates-of-strain* that exist in a fluid body, as considered in the previous chapter, and as in that case they are empirical in origin.

If the strains in the axial, radial, and angular directions are denoted by $e_{xx}, e_{rr}, e_{\theta\theta}$, the stress–strain relations for an elastic body are given by

$$e_{xx} = \frac{1}{E}\left[S_{xx} - \sigma(S_{rr} + S_{\theta\theta})\right] \tag{5.4.1}$$

$$e_{rr} = \frac{1}{E}\left[S_{rr} - \sigma(S_{\theta\theta} + S_{xx})\right] \tag{5.4.2}$$

$$e_{\theta\theta} = \frac{1}{E}\left[S_{\theta\theta} - \sigma(S_{rr} + S_{xx})\right] \tag{5.4.3}$$

where E, σ are two constant properties of the elastic material, known as Young's modulus and Poisson's ratio, respectively. The relations express a fundamental characteristic of elastic materials whereby the strain in one direction depends not only on stress in that direction but also on stresses in the other two directions.

Using the relation between the radial and angular stresses (Eq.5.3.12), only two of the above relations are required, namely,

$$e_{xx} = \frac{1}{E}\left[S_{xx} - \sigma S_{rr}\left(1 + \frac{a}{h}\right)\right] \tag{5.4.4}$$

$$e_{\theta\theta} = \frac{1}{E}\left[S_{rr}\left(\frac{a}{h} - \sigma\right) - \sigma S_{xx}\right] \tag{5.4.5}$$

Assuming that a/h is sufficiently large compared with 1.0 or σ, these reduce to

$$e_{xx} = \frac{1}{E}\left[S_{xx} - \frac{\sigma a}{h} S_{rr}\right] \tag{5.4.6}$$

$$e_{\theta\theta} = \frac{1}{E}\left[\frac{a}{h} S_{rr} - \sigma S_{xx}\right] \tag{5.4.7}$$

and solving for the two stresses, we obtain finally

$$S_{xx} = E_\sigma(e_{xx} + \sigma e_{\theta\theta}) \tag{5.4.8}$$

$$S_{rr} = \frac{hE_\sigma}{a}(e_{\theta\theta} + \sigma e_{xx}) \tag{5.4.9}$$

where

$$E_\sigma = \frac{E}{1 - \sigma^2} \tag{5.4.10}$$

Axial strain is caused by elongation of the tube, which in turn is caused by variation of the axial displacement ξ along the tube, that is, by ξ being a function of x. If all elements of the tube undergo the *same* axial displacement, that is, if ξ is constant, the axial strain is zero. More generally ξ is a function of x and a small element of the wall of original length δx will have length

$$\delta x + \delta\xi = \delta x + \frac{\partial \xi}{\partial x}\delta x \tag{5.4.11}$$

in its strained state (Fig.5.4.1). Axial strain is defined as the ratio of the change in length over original length, that is

$$\begin{aligned} e_{xx} &= \frac{1}{\delta x}\left[\delta x - \left(\delta x + \frac{\partial \xi}{\partial x}\delta x\right)\right] \\ &= \frac{\partial \xi}{\partial x} \end{aligned} \tag{5.4.12}$$

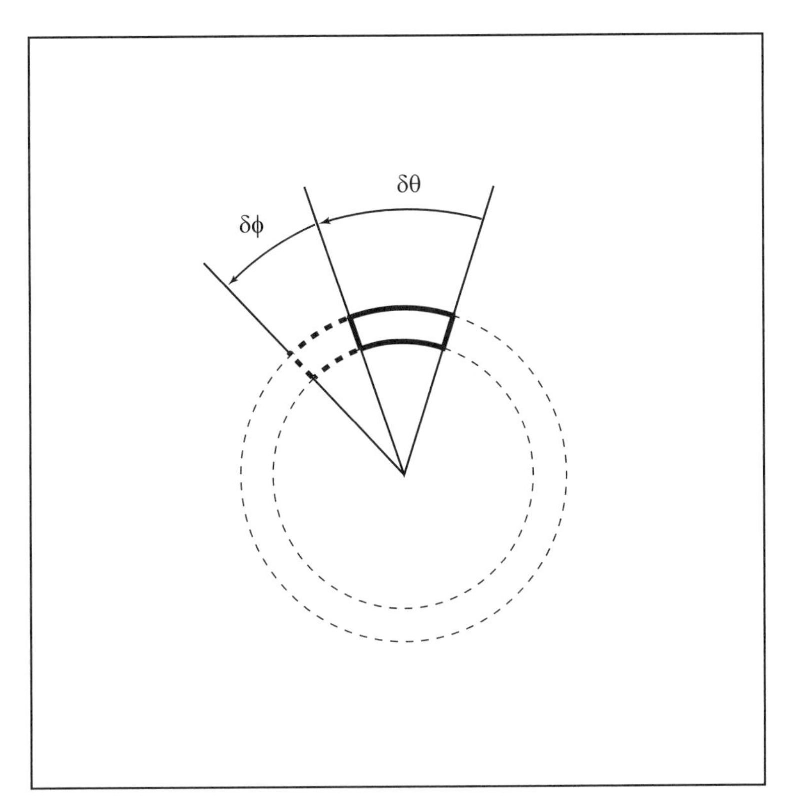

FIGURE 5.4.2. Angular displacement ϕ may in general be different at different points around the tube wall, leading to angular elongation. On the assumption of axial symmetry, however, angular displacement ϕ is uniform around the tube, that is, $\delta\phi = 0$, and hence this source of angular elongation is zero. A more important source of angular strain, which does not conflict with axial symmetry, is shown in Fig.5.4.3.

Angular strain may arise in two ways. First in analogy with axial strain, a segment of the tube subtended by angle $\delta\theta$ and of original length $a\delta\theta$ may change its length because the angular displacement ϕ is not the same all around the tube, that is, because of a gradient $\partial\phi/\partial\theta$ (Fig.5.4.2). Because of axial symmetry, however, this gradient is here assumed to be zero and hence this source of angular strain is zero. Another, more important, source of angular strain is *radial* displacement η, which changes the radius of the tube from its neutral radius a to $a+\eta$ and therefore changes the arc length of the segment from $a\delta\theta$ to $(a+\eta)\delta\theta$ (Fig.5.4.3). Angular strain is defined as the ratio of change in length over original length, that is,

$$e_{\theta\theta} = \frac{1}{a\delta\theta}[(a+\eta)\delta\theta - a\delta\theta]$$

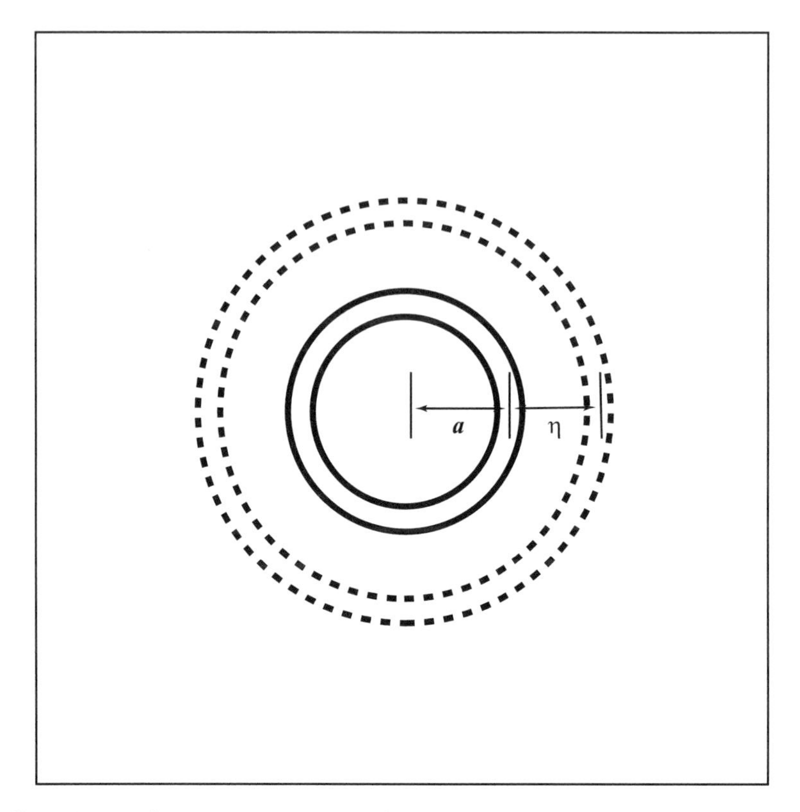

FIGURE 5.4.3. An important source of angular strain is a change in tube radius, as shown, from a to $a+\eta$. The resulting *angular* strain, as discussed in the text, is η/a (Eq.5.4.13).

$$= \frac{\eta}{a} \tag{5.4.13}$$

Substituting these results in Eqs.5.4.8,9, the expressions for the axial and radial stresses become

$$S_{xx} = E_\sigma \left(\frac{\partial \xi}{\partial x} + \sigma \frac{\eta}{a} \right) \tag{5.4.14}$$

$$S_{rr} = \frac{hE_\sigma}{a} \left(\frac{\eta}{a} + \sigma \frac{\partial \xi}{\partial x} \right) \tag{5.4.15}$$

and substituting these in turn into Eqs.5.3.8,10, the equations of motion of the tube wall become

$$\frac{\partial^2 \xi}{\partial t^2} = \frac{E_\sigma}{\rho_w} \left(\frac{\partial^2 \xi}{\partial x^2} + \frac{\sigma}{a} \frac{\partial \eta}{\partial x} \right) - \frac{\tau_w}{\rho_w h} \tag{5.4.16}$$

$$\frac{\partial^2 \eta}{\partial t^2} = \frac{p_w}{\rho_w h} - \frac{E_\sigma}{\rho_w a} \left(\frac{\eta}{a} + \sigma \frac{\partial \xi}{\partial x} \right) \tag{5.4.17}$$

The equations are coupled with those of the flow field through the pressure p_w and shear stress τ_w and this coupling is dealt with in the next section.

5.5 Coupling with Fluid Motion

Motion of the tube wall is coupled to the motion of the fluid through the action of fluid pressure and shear stress on the tube wall as illustrated in Fig.5.5.1. Mathematically, the coupling occurs through the presence of p_w and τ_w in the equations of wall motion (Eqs.5.4.16,17). To solve the equations these flow parameters must be determined from the flow field solution, which we do in this section.

The pressure acting on the tube wall, from Eqs.5.2.1,20, is given by

$$p_w = p(x,a,t) = Be^{i\omega(t-x/c)} \tag{5.5.1}$$

The shear stress acting on the tube wall, from Eqs.2.7.2, 3.4.6, is given by

$$\tau_w = -(\tau_{rx})_{r=a} = -\mu\left(\frac{\partial u}{\partial r} + \frac{\partial v}{\partial x}\right)_{r=a} \tag{5.5.2}$$

Applying the approximations used before, namely, that the length of the traveling wave is much larger than the tube radius, the second gradient above is much smaller than the first and can be neglected, so that we take

$$\tau_w = -\mu\left(\frac{\partial u}{\partial r}\right)_{r=a} \tag{5.5.3}$$

Using the result for the velocity in Eq.5.2.2, we have

$$\tau_w = -\mu\left(\frac{dU(r)}{dr}\right)_{r=a} e^{i\omega(t-x/c)} \tag{5.5.4}$$

and using the the result for $U(r)$ in Eq.5.2.11, this gives, after some algebra

$$\tau_w = \left(-\frac{\mu A\Lambda J_1(\Lambda)}{a} + \frac{\mu B\omega^2 a}{2\rho c^3}\right) e^{i\omega(t-x/c)} \tag{5.5.5}$$

where we have again used the approximations used in the flow field (Eqs.5.2.15–17). Substituting for p_w and τ_w, the equations of wall motion become

$$\begin{aligned}\frac{\partial^2\xi}{\partial t^2} &= \frac{E_\sigma}{\rho_w}\left(\frac{\partial^2\xi}{\partial x^2} + \frac{\sigma}{a}\frac{\partial\eta}{\partial x}\right)\\ &- \frac{1}{\rho_w h}\left(-\frac{\mu A\Lambda J_1(\Lambda)}{a} + \frac{\mu B\omega^2 a}{2\rho c^3}\right) e^{i\omega(t-x/c)}\end{aligned} \tag{5.5.6}$$

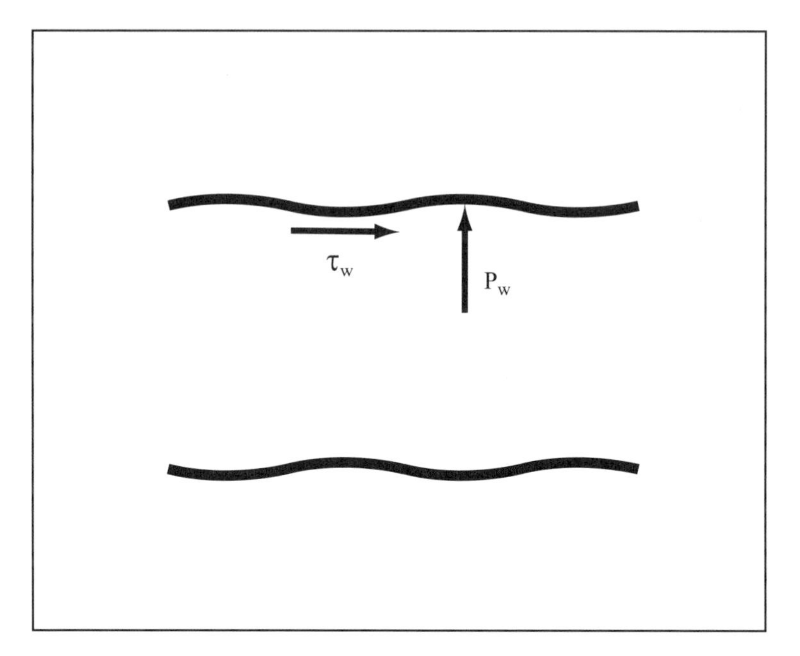

FIGURE 5.5.1. Motion of the tube wall is mediated by two stresses exerted by the moving fluid on the inner surface of the tube wall: pressure p_w and shear stress τ_w.

$$\frac{\partial^2 \eta}{\partial t^2} = \frac{B}{\rho_w h} e^{i\omega(t-x/c)} - \frac{E_\sigma}{\rho_w a}\left(\frac{\eta}{a} + \sigma\frac{\partial \xi}{\partial x}\right) \tag{5.5.7}$$

5.6 Matching at the Tube Wall

The equations of wall motion now contain two arbitrary constants A, B yet to be determined. These constants provide the link between motion of the fluid and motion of the tube wall, and are determined by matching the two motions at the interface between the fluid and the inner surface of the tube wall. The matching is expressed in terms of two boundary conditions at the tube wall, which require that the radial and axial velocities of the wall be equal to the radial and axial velocities of the fluid in contact with the wall. As before, because the wall is itself in motion, these boundary conditions are applied only approximately at the neutral position of the wall, $r = a$, that is, we take

$$\frac{\partial \xi}{\partial t} = u(x, a, t) \tag{5.6.1}$$

$$\frac{\partial \eta}{\partial t} = v(x, a, t) \tag{5.6.2}$$

It is reasonable to assume that the axial and radial oscillatory movements of the wall have the same frequency as that prevailing in the flow field, thus we write

$$\xi(x,t) = Ce^{i\omega(t-x/c)} \tag{5.6.3}$$
$$\eta(x,t) = De^{i\omega(t-x/c)} \tag{5.6.4}$$

where C, D are two new constants to be determined. Note that this form does not imply that the wall motion is *in phase* with the oscillatory motion of the fluid, because these constants, as we shall find out, are generally complex quantities. Substituting for ξ, η and their derivatives in the two equations of wall motion (Eqs.5.4.16,17) and two boundary conditions (Eqs.5.6.1,2), we obtain a set of four equations for the constants A, B, C, D, namely,

$$\begin{aligned} -\omega^2 C &= \frac{E_\sigma}{\rho_w}\left[-\frac{\omega^2}{c^2}C + \frac{\sigma}{a}\left(\frac{-i\omega}{c}\right)D\right] \\ &\quad - \frac{1}{\rho_w h}\left[-\frac{\mu\Lambda J_1(\Lambda)}{a}A + \frac{\mu\omega^2 a}{2\rho c^3}B\right] \end{aligned} \tag{5.6.5}$$

$$-\omega^2 D = \frac{B}{\rho_w h} - \frac{E_\sigma}{\rho_w a}\left[\frac{D}{a} + \sigma\left(\frac{-i\omega}{c}\right)C\right] \tag{5.6.6}$$

$$i\omega C = J_0(\Lambda)A + \frac{B}{\rho c} \tag{5.6.7}$$

$$i\omega D = \frac{i\omega a J_1(\Lambda)}{c\Lambda}A + \frac{i\omega a}{2\rho c^2}B \tag{5.6.8}$$

Some simplification is possible by noting that in the first equation

$$\begin{aligned} -\frac{1}{\rho_w h} &\times \left(-\frac{\mu\Lambda J_1(\Lambda)}{a}A + \frac{\mu\omega^2 a}{2\rho c^3}B\right) \\ &= \frac{\mu}{\rho_w hca}\left(\Lambda J_1(\Lambda)cA - \frac{\omega^2 a^2}{2c^2}B\right) \\ &\approx \frac{\mu J_1(\Lambda)\Lambda}{\rho_w ha}A \end{aligned} \tag{5.6.9}$$

and in the second equation

$$\begin{aligned} -\omega^2 D + \frac{E_\sigma}{\rho_w a^2}D &= \frac{c^2}{a^2}\left(-\frac{\omega^2 a^2}{c^2} + \frac{E_\sigma}{\rho_w c^2}\right)D \\ &\approx \frac{E_\sigma}{\rho_w a^2}D \end{aligned} \tag{5.6.10}$$

In both cases the quantity $\omega^2 a^2/c^2$ is being neglected as it is of order $(a/L)^2$, where a is tube radius and L is wave length of the propagating wave.

With these simplifications, the four combined equations for A, B, C, D take the final form

$$-\omega^2 C = \frac{E_\sigma}{\rho_w}\left[-\frac{\omega^2}{c^2}C + \frac{\sigma}{a}\left(\frac{-i\omega}{c}\right)D\right] + \left[\frac{\mu\Lambda J_1(\Lambda)}{\rho_w h a}\right]A \tag{5.6.11}$$

$$0 = \frac{B}{\rho_w h} - \frac{E_\sigma}{\rho_w a}\left[\frac{D}{a} + \sigma\left(\frac{-i\omega}{c}\right)C\right] \tag{5.6.12}$$

$$i\omega C = J_0(\Lambda)A + \frac{B}{\rho c} \tag{5.6.13}$$

$$i\omega D = \frac{i\omega a J_1(\Lambda)}{c\Lambda}A + \frac{i\omega a}{2\rho c^2}B \tag{5.6.14}$$

5.7 Wave Speed

The two equations of wall motion and two boundary conditions obtained in the previous section provide a set of four equations for the four remaining unknown arbitrary constants A, B, C, D. Another important unknown that remains to be determined is the wave speed c (not to be confused with the constant C). In this section we shall see how this speed is determined, and in the next section we deal with the four arbitrary constants.

Equations 5.6.11–14 can be put in the form of the following four linear equations in A, B, C, D

$$a_{11}A + a_{13}C + a_{14}D = 0 \tag{5.7.1}$$
$$a_{22}B + a_{23}C + a_{24}D = 0 \tag{5.7.2}$$
$$a_{31}A + a_{32}B + a_{33}C = 0 \tag{5.7.3}$$
$$a_{41}A + a_{42}B + a_{44}D = 0 \tag{5.7.4}$$

where the coefficients are given by

$$a_{11} = \frac{\mu\Lambda J_1(\Lambda)}{\rho_w h a} \tag{5.7.5}$$

$$a_{13} = \omega^2\left(1 - \frac{E_\sigma}{\rho_w c^2}\right) \tag{5.7.6}$$

$$a_{14} = \frac{-i\omega\sigma E_\sigma}{\rho_w a c} \tag{5.7.7}$$

$$a_{22} = \frac{1}{h} \tag{5.7.8}$$

$$a_{23} = \frac{i\omega\sigma E_\sigma}{ac} \tag{5.7.9}$$

$$a_{24} = \frac{-E_\sigma}{a^2} \tag{5.7.10}$$

$$a_{31} = J_0(\Lambda) \tag{5.7.11}$$

$$a_{32} = \frac{1}{\rho c} \tag{5.7.12}$$

$$a_{33} = -i\omega \tag{5.7.13}$$

$$a_{41} = \frac{i\omega J_1(\Lambda)a}{c\Lambda} \tag{5.7.14}$$

$$a_{42} = \frac{i\omega a}{2\rho c^2} \tag{5.7.15}$$

$$a_{44} = -i\omega \tag{5.7.16}$$

Because the system of four equations is *homogeneous*, a nontrivial solution is obtained only if the determinant of the coefficients is zero [12–14], that is, if

$$\begin{vmatrix} a_{11} & 0 & a_{13} & a_{14} \\ 0 & a_{22} & a_{23} & a_{24} \\ a_{31} & a_{32} & a_{33} & 0 \\ a_{41} & a_{42} & 0 & a_{44} \end{vmatrix} = 0$$

or

$$\begin{aligned} & a_{11}[a_{22}(a_{33}a_{44}) - a_{23}(a_{32}a_{44}) + a_{24}(-a_{42}a_{33})] \\ & \quad +a_{13}[-a_{22}(a_{31}a_{44}) + a_{24}(a_{31}a_{42} - a_{41}a_{32})] \\ & -a_{14}[-a_{22}(-a_{41}a_{33}) + a_{23}(a_{31}a_{42} - a_{41}a_{32})] = 0 \end{aligned} \tag{5.7.17}$$

Substituting for the coefficients, this gives, after some algebra,

$$\begin{aligned} [(g-1)(\sigma^2-1)]z^2 & + \left[\frac{\rho_w h}{\rho a}(g-1) + \left(2\sigma - \frac{1}{2}\right) g - 2\right] z \\ & + \frac{2\rho_w h}{\rho a} + g = 0 \end{aligned} \tag{5.7.18}$$

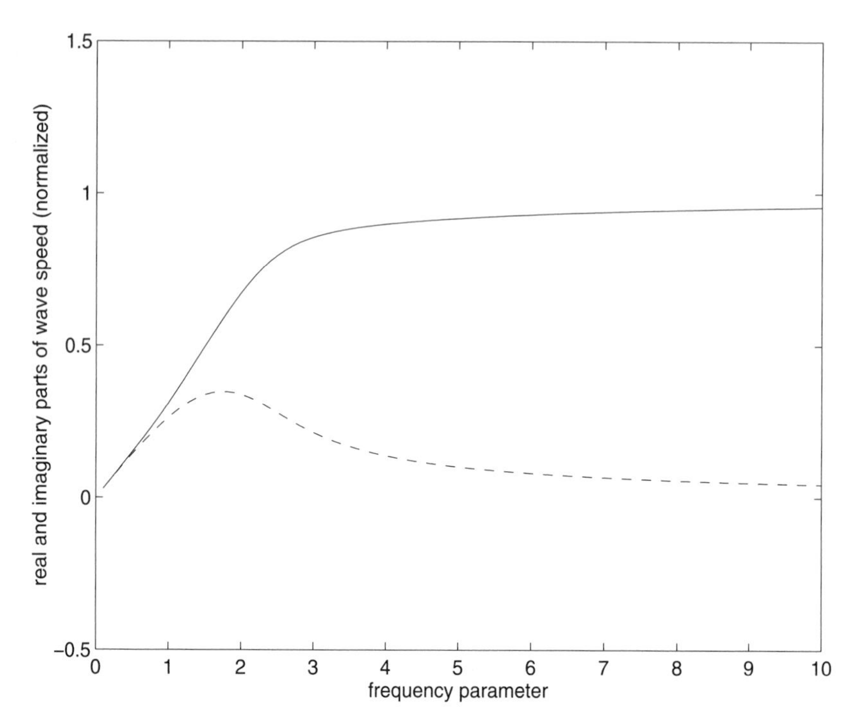

FIGURE 5.7.1. Variation of the real (solid line) and imaginary (dashed line) parts of the wave speed c, normalized in terms of the wave speed in invisicid flow, c_0, with frequency parameter Ω. As the frequency increases, the imaginary part of c vanishes while the real part becomes the same as c_0.

where

$$z = \frac{E_\sigma h}{\rho a c^2} \tag{5.7.19}$$

$$g = \frac{2J_1(\Lambda)}{\Lambda J_0(\Lambda)} \tag{5.7.20}$$

Equation 5.7.18 is a quadratic equation in z and its solution therefore furnishes a value of the wave speed c in terms of parameters of the fluid and tube wall. In particular, recalling that the wave speed in *inviscid* flow is given by (Eq.5.1.1)

$$c_0^2 = \frac{Eh}{2\rho a}$$

then

$$z = \frac{E_\sigma h}{\rho a c^2} = \frac{2}{1-\sigma^2}\left(\frac{c_0}{c}\right)^2$$

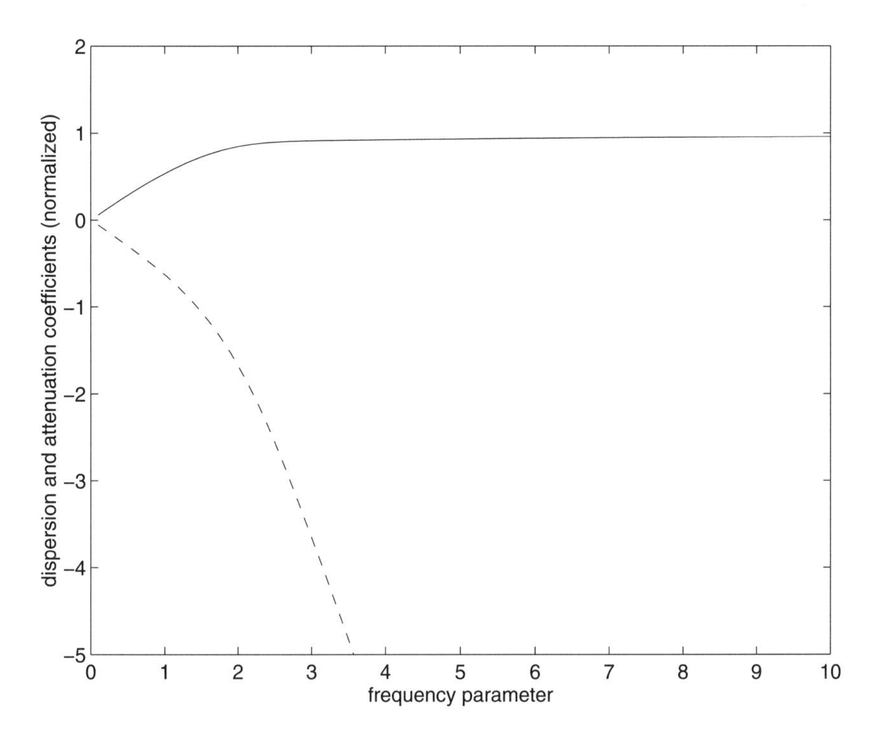

FIGURE 5.7.2. Variation of dispersion and attenuation coefficients with frequency parameter Ω. Solid and dashed lines show c_1 and c_2 respectively, as defined by Eq.5.7.22, and normalized in terms of the wave speed in inviscid flow, c_0. As the frequency increases, $c_2/c_0 \to -\infty$ and $c_1/c_0 \to 1.0$, and both attenuation and dispersion effects vanish from Eq.5.7.23.

$$c = \sqrt{\frac{2}{(1-\sigma^2)z}}\, c_0 \qquad (5.7.21)$$

Thus z is a measure of the wave speed in viscous flow as compared with that in inviscid flow.

Because z is complex, it follows that the wave speed c is also complex and is therefore not a true "speed." This is to be contrasted with wave propagation in *inviscid* flow where the wave speed c_0 is real and is a true speed in the physical sense. Furthermore, while c_0 depends on constant properties of the tube and the fluid only, c depends also on frequency because the solution for z depends on frequency.

To examine the consequences of this, it is convenient to write

$$\frac{1}{c} = \frac{1}{c_1} + i\frac{1}{c_2} \qquad (5.7.22)$$

so that

$$e^{i\omega(t-x/c)} = e^{i\omega(t-x/c_1-ix/c_2)}$$

$$= e^{\omega x/c_2} e^{i\omega(t-x/c_1)} \tag{5.7.23}$$

Comparing this waveform with that in *inviscid* flow, where $c = c_0$, we have, for the inviscid case

$$e^{i\omega(t-x/c)} = e^{i\omega(t-x/c_0)} \tag{5.7.24}$$

It is seen that the effect of viscosity is to change the amplitude of the wave from a reference value of 1.0 in the inviscid case to $e^{\omega x/c_2}$ in the viscous case, an effect usually referred to as "attenuation." Another effect of viscosity is to change the wave speed from c_0 in the inviscid case to c_1 in the viscous case (Eqs.5.7.23,24). Because c_1 depends on frequency, this effect will be different for the different harmonic components of a composite wave, an effect known as "dispersion." The dependence of c_1 on frequency is therefore an important element of the solution of the problem of pulsatile flow in an elastic tube.

Solution of Eq.5.7.18 provides a value of c for each value of the frequency, hence it is sometimes referred to as the "frequency equation." Results, in Fig.5.7.1, show how the real and imaginary parts of c vary with the frequency parameter Ω. Results for c_1 and c_2, sometimes referred to as the dispersion and attenuation coefficients, respectively, are shown in Fig.5.7.2.

5.8 Arbitrary Constants

The four equations for the constants A, B, C, D are a set of homogeneous linear equations, with a 4×4 coefficient matrix which is of rank 3, therefore one of A, B, C, D must remain arbitrary [12–14]. That is, the solution determines only three of A, B, C, D in terms of the fourth. Results of the flow field solution suggest that the obvious choice to make is that of expressing A, C, D in terms of B, since the latter is equal to the amplitude of the input oscillatory pressure which would normally be known or specified. Thus from the flow field solution (Eqs.5.2.1,20) we have

$$\begin{aligned} p(x,r,t) &= P(r)e^{i\omega(t-x/c)} \\ &= Be^{i\omega(t-x/c)} \end{aligned} \tag{5.8.1}$$

which in effect determines that $P(r)$ is constant and is equal to B. Thus our solution of Eqs.5.7.1–4 for the arbitrary constants begins with B assumed known, representing in fact the amplitude of the input oscillatory pressure at $x = 0$.

Using Eqs.5.7.2,4 to eliminate D, and combining the result with Eq.5.7.3 gives

$$A = \frac{a_{33}a_{24}a_{42} - a_{33}a_{44}a_{22} + a_{44}a_{23}a_{32}}{-a_{33}a_{24}a_{41} - a_{44}a_{23}a_{31}} B \tag{5.8.2}$$

From Eq.5.7.3 we then have

$$C = \frac{a_{31}A + a_{32}B}{-a_{33}} \tag{5.8.3}$$

and from Eq.5.7.4

$$D = \frac{a_{41}A + a_{42}B}{-a_{44}} \tag{5.8.4}$$

Substituting for the coefficients from Eqs.5.7.5–16, we get, finally,

$$A = \frac{1}{\rho c J_0(\Lambda)} \left[\frac{2 + z(2\sigma - 1)}{z(g - 2\sigma)} \right] B \tag{5.8.5}$$

$$C = \frac{i}{\rho c \omega} \left[\frac{2 - z(1 - g)}{z(2\sigma - g)} \right] B \tag{5.8.6}$$

$$D = \frac{a}{\rho c^2} \left[\frac{g + \sigma z(g - 1)}{z(g - 2\sigma)} \right] B \tag{5.8.7}$$

where z, g are defined by Eqs.5.7.19,20.

5.9 Flow Properties

The axial velocity is now fully determined, using Eqs.5.2.2,18, 5.8.5,

$$u(x, r, t) = \frac{B}{\rho c} \left[1 - G \frac{J_0(\zeta)}{J_0(\Lambda)} \right] e^{i\omega(t - x/c)} \tag{5.9.1}$$

where G shall be referred to as the "elasticity factor", given by

$$G = \frac{2 + z(2\sigma - 1)}{z(2\sigma - g)} \tag{5.9.2}$$

This is a classical solution of the problem of oscillatory flow in an elastic tube, obtained by Morgan and Kiely [15] and Womersley [16] and enlarged upon by others [17–19], although the rudiments of the solution can be traced back to pioneering work by Korteweg [20], Lamb [21], Witzig [22], and Lambossy [23].

To compare this and other results with corresponding properties of *steady* flow in a tube, we recall (Eq.5.8.1) that the constant B in the present case represents the amplitude of the input oscillatory pressure, which must be specified for the problem to be complete. For comparison we take the amplitude of the oscillatory pressure *gradient* to be equal to the constant pressure gradient in steady flow, namely, k_s as defined in Eq.3.3.3. Using Eq.5.8.1, then

$$p(x, r, t) = B e^{i\omega(t - x/c)}$$

$$\frac{\partial p}{\partial x} = -\frac{i\omega}{c}Be^{i\omega(t-x/c)} \tag{5.9.3}$$

and we take

$$\frac{-i\omega}{c}B = k_s, \qquad \text{that is,} \quad B = \frac{ic}{\omega}k_s \tag{5.9.4}$$

It is useful also to nondimensionalize the axial velocity in oscillatory flow in terms of the maximum velocity in steady flow, namely, from Eq.3.4.1,

$$\hat{u}_s = -\frac{k_s a^2}{4\mu}$$

thus we finally have

$$\frac{u(x,r,t)}{\hat{u}_s} = \frac{-4}{\Lambda^2}\left[1 - G\frac{J_0(\zeta)}{J_0(\Lambda)}\right]e^{i\omega(t-x/c)} \tag{5.9.5}$$

Comparison with the corresponding expression for pulsatile flow in a *rigid* tube (Eq.4.6.2) indicates that the difference between the two is contained entirely in the value of the elasticity factor G. However, since G is complex and both its real and imaginary parts depend on the frequency ω, the effect is not easily apparent. Variation of the real and imaginary parts of G with frequency are shown in Fig.5.9.1. Its effects on the oscillatory velocity profiles are shown in Figs.5.9.2–4, where the profiles are compared with the corresponding velocity profiles of oscillatory flow in a rigid tube.

In the interpretation of Figs.5.9.2–4 it is important to note that oscillatory flow in an elastic tube consists of *two* oscillations, one in time and one in space. Both have the same frequency ω and hence the same period $T = 2\pi/\omega$. During this time period, the input oscillatory pressure completes one cycle, in time, while the pressure within the tube completes one cycle *in space*, which is the essence of wave propagation. The length of tube which this cycle occupies is the wavelength $L = cT = 2\pi c/\omega$. Over this length of tube, the pressure completes one back-and-forth oscillation while the input pressure completes one back-and-forth oscillation in time.

In Figs.5.9.2–4, the five panels in each figure represent the back-and-forth oscillations in time, while the four profiles in each panel represent the back-and-forth oscillations in space, over a length of tube equal to the wavelength L. Thus the five velocity profiles shown in the top panel represent velocity profiles *at a fixed point in time*, $t = 0$, *within the oscillatory cycle* but prevailing over one wavelength L along the tube. The profiles are equally spaced at quarter wave length intervals, so that $L = 4.0$ on the scale of the figure. At each subsequent panel the corresponding picture is shown at a later time in the oscillatory cycle, namely, at $t = T/4$, $2T/4$, $3T/4$, and $4T/4$. The three figures represent this complete picture at low, medium, and high frequency, $\Omega = 1, 3, 10$.

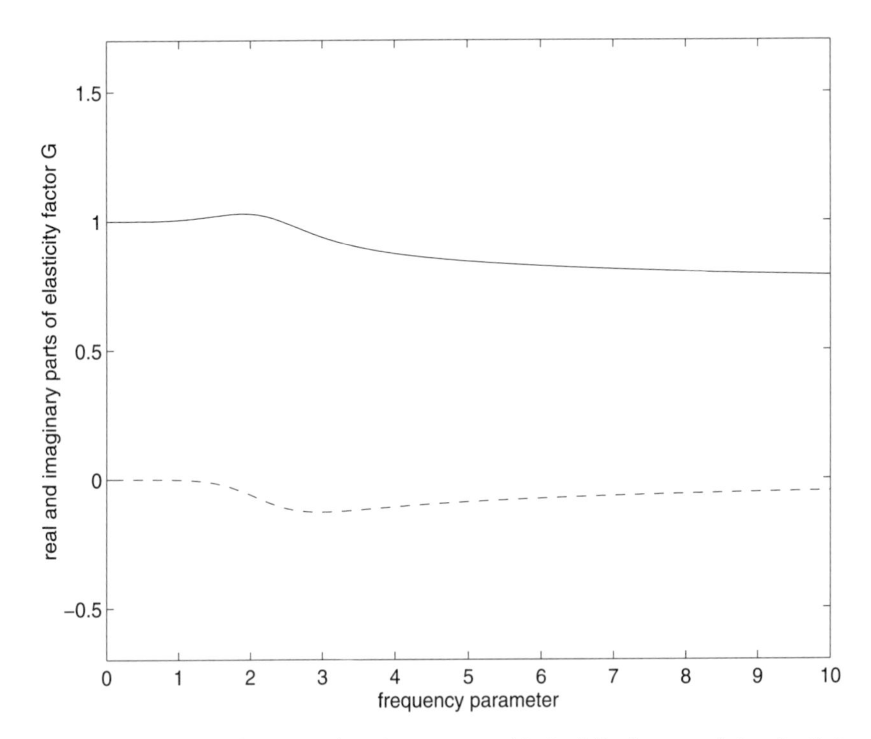

FIGURE 5.9.1. Real (solid line) and imaginary (dashed line) parts of the elasticity factor G, which embodies the difference between oscillatory flow in rigid and elastic tubes (see text).

The large differences seen in Figs.5.9.2–4 between flow in elastic and rigid tubes therefore require a complete wavelength L within the tube to be observed. In a tube shorter than L, the differences will not occur in full. Tube lengths l in the vascular system range, in orders of magnitude, from $l = L$ to $l = L/100$, and results for $L/10$ and $L/100$ are therefore shown in Figs.5.9.5,6, for comparison with those in Fig.5.9.2. The results for $l = L/100$ show clearly that the difference between flow in rigid and elastic tubes becomes insignificantly small in the limit of small l/L.

The radial velocity, using Eqs.5.2.19, 5.8.5, and nondimensionalizing in terms of the maximum velocity in steady flow, is given by

$$\frac{v(x,r,t)}{\hat{u}_s} = \frac{2a\omega}{i\Lambda^2 c}\left[\frac{r}{a} - G\frac{2J_1(\zeta)}{\Lambda J_0(\Lambda)}\right] e^{i\omega(t-x/c)} \tag{5.9.6}$$

At the tube wall, $r = a$, this becomes

$$\frac{v(x,a,t)}{\hat{u}_s} = \frac{2a\omega}{i\Lambda^2 c}[1 - Gg]e^{i\omega(t-x/c)} \tag{5.9.7}$$

which equals the radial velocity of the tube wall and is therefore of particular interest. Variation of the radial velocity over one wavelength along

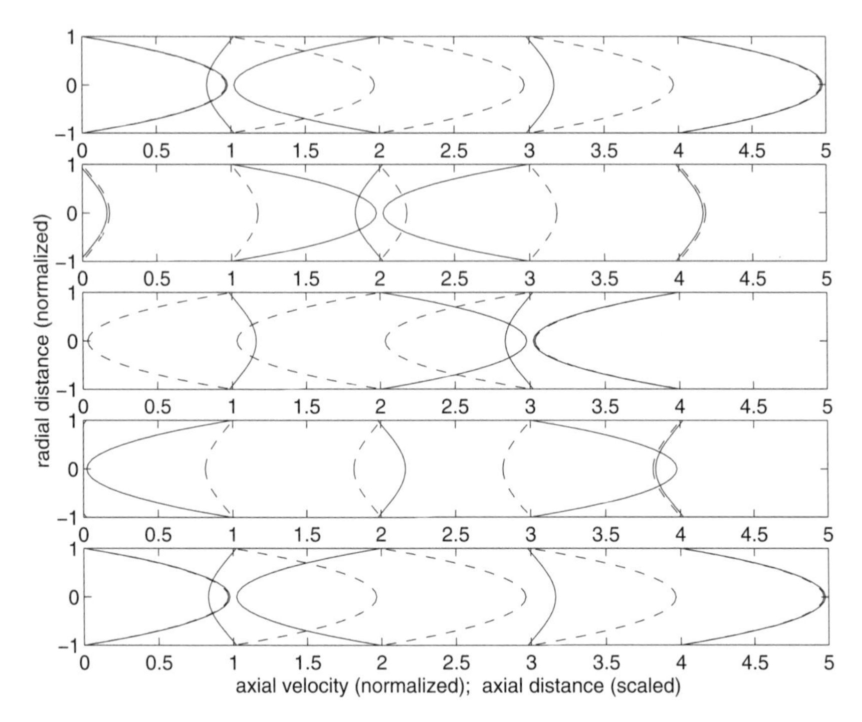

FIGURE 5.9.2. Oscillatory velocity profiles in an elastic tube (solid lines) compared with those in a rigid tube (dashed lines). Oscillatory flow in an elastic tube consists of *two* oscillations, one in time and one in space. Both have the same frequency ω and hence the same period $T = 2\pi/\omega$. During this time period, the input oscillatory pressure completes one cycle, in time, while the pressure within the tube completes one cycle *in space* over a length of tube $L = cT = 2\pi c/\omega$. In this figure the five panels represent the back-and-forth oscillations in time while the five profiles in each panel represent the back-and-forth oscillations in space, over a length of tube equal to the wavelength L. Thus the four velocity profiles shown in the top panel represent velocity profiles *at a fixed point in time, $t = 0$, within the oscillatory cycle* but prevailing over one wavelength L along the tube. The profiles are equally spaced at quarter wavelength intervals, so that $L = 4.0$ on the scale of the figure. At each subsequent panel the corresponding picture is shown at a later time in the oscillatory cycle, namely, at $t = T/4$, $2T/4$, $3T/4$, and $4T/4$. Results are for low frequency, $\Omega = 1.0$.

the tube are shown in Fig.5.9.7 at low frequency ($\Omega = 1.0$). The velocity is scaled by the factor $2a\omega/i\Lambda^2 c$, so that the quantity shown is only $[1 - Gg]e^{i\omega(t-x/c)}$. The results mimic the movement of the tube wall within the oscillatory cycle, and over one wavelength L along the tube.

The flow rate

$$q(x,t) = \int_0^a 2\pi r u dr \tag{5.9.8}$$

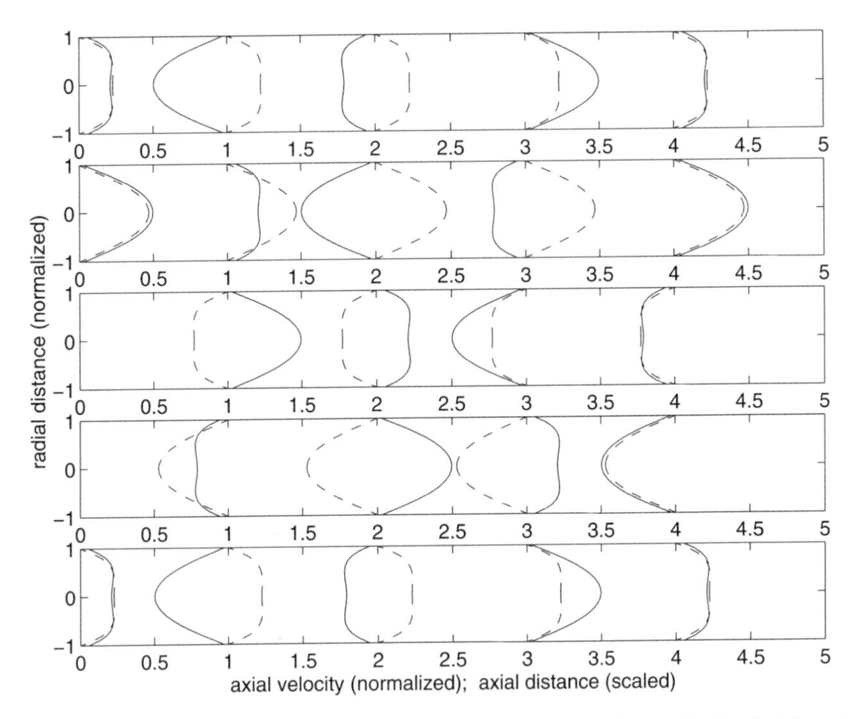

FIGURE 5.9.3. Oscillatory velocity profiles in an elastic tube, as in Fig.5.9.2, but at moderate frequency, $\Omega = 3.0$.

depends on the radius of the tube, which here is not constant as it is in the case of pulsatile flow in a rigid tube. On the assumption that the radial movements are small, an approximate value of the flow rate is obtained by treating a as a constant neutral radius of the tube. Nondimensionalizing in terms of the flow rate in steady flow, q_s in Eq.3.4.3, this gives

$$\frac{q(x,t)}{q_s} = \frac{-8}{\Lambda^2}(1 - Gg)e^{i\omega(t-x/c)} \tag{5.9.9}$$

Results for oscillatory flow rate at moderate frequency ($\Omega = 3.0$) are shown in Fig.5.9.8, compared with the corresponding flow rate in a rigid tube. It is seen that a small difference exists between the two cases at this frequency, with the flow rate in an elastic tube reaching a somewhat higher peak than that in a rigid tube. The percentage difference between the two peaks at different frequencies is shown in Fig.5.9.9. Wall movements in an elastic tube make it somewhat "easier" for the flow to move through the tube.

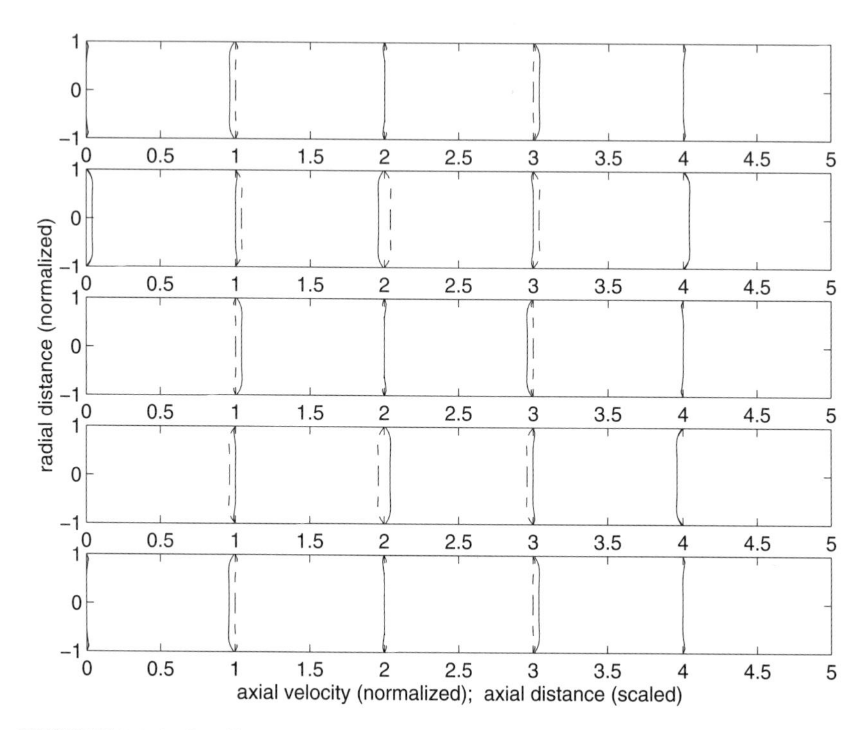

FIGURE 5.9.4. Oscillatory velocity profiles in an elastic tube, as in Fig.5.9.2, but at high frequency, $\Omega = 10.0$.

5.10 Problems

1. The simplified equations governing pulsatile flow in an elastic tube (Eqs.5.1.4–6) are based on two key assumptions in addition to those on which the equation governing pulsatile flow in a rigid tube is based. What are these?
2. Are the assumptions for pulsatile flow in an elastic tube better satisfied in a "stiffer" tube or one that is more elastic? Are they better satisfied in a tube of smaller or larger radius?
3. What is the wave speed in the case of pulsatile flow in a *rigid* tube? Is the question a legitimate one?
4. In pulsatile flow through an elastic tube the radius of the tube is no longer a constant, in fact, it is a function of both position x along the tube and time t. Yet in the basic solution of the flow equations (Eqs.5.2.4–6) it was assumed that boundary conditions can be applied at $r = a$ (Eqs.5.2.10). Discuss the physical basis and implications of this assumption.
5. Describe in physical terms the forces acting on a segment of the tube wall in pulsatile flow through an elastic tube.

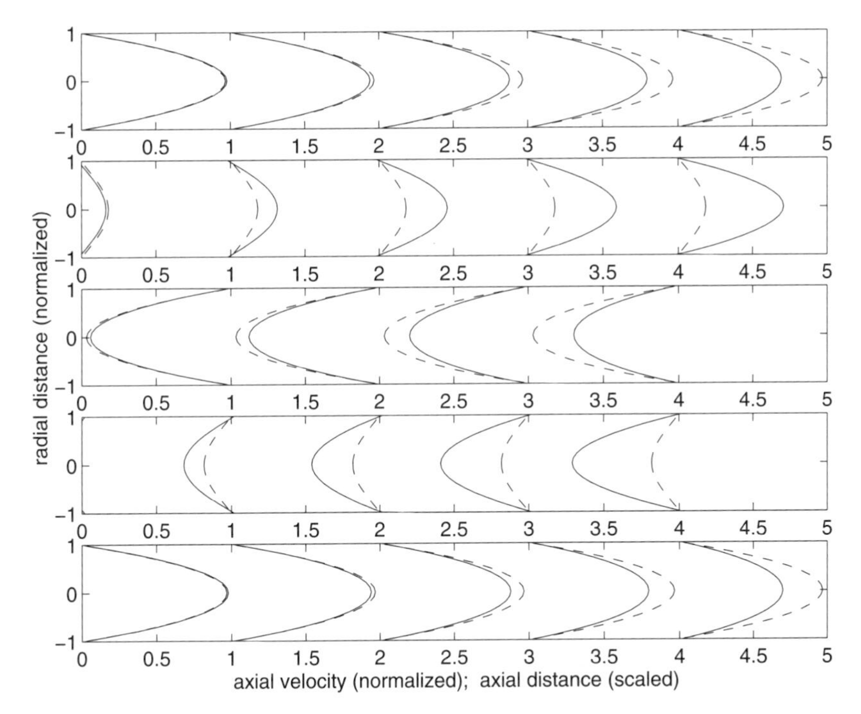

FIGURE 5.9.5. Oscillatory velocity profiles in an elastic tube, at low frequency, $\Omega = 1.0$, as in Fig.5.9.2, but here in a tube of length $l = L/10$ where L is wavelength. In Fig.5.9.2, by comparison, the tube length and wavelength are equal.

6. Describe in physical terms the main elements of Eqs.5.4.16,17 governing the motion of a segment of the tube wall in pulsatile flow through an elastic tube.
7. The assumption that the tube wall thickness is much smaller than the tube radius is central in the analysis of pulsatile flow in an elastic tube. Indicate the points in the analysis at which the assumption is invoked, and discuss the physical implications in each case.
8. Describe the nature of analytical coupling between the equations governing fluid flow and those governing wall movement in pulsatile flow in an elastic tube.
9. Describe in physical terms the matching conditions that must be applied to resolve the coupling between the tube wall movement and the flow field within the tube.
10. Describe in physical terms the difference between the wave speed c_0 as defined in Eq.5.1.1, and the wave speed c obtained from a solution of the coupled equations for pulsatile flow in an elastic tube (Eq.5.7.21).
11. From Eq.5.9.9 for flow rate, and using tables in Appendix A, calculate the percentage difference between the peak value of oscillatory flow rate

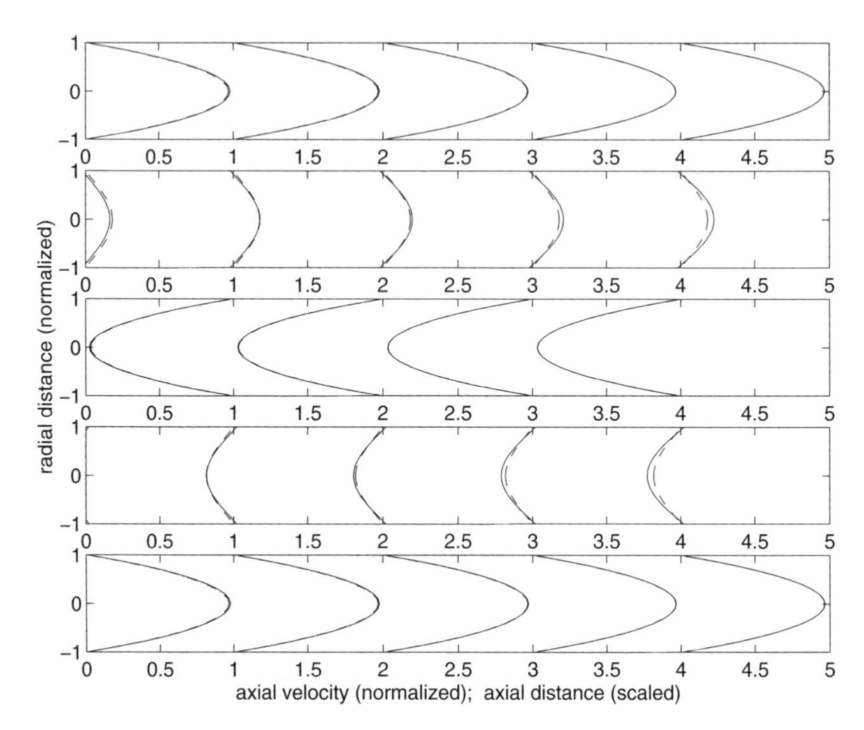

FIGURE 5.9.6. Oscillatory velocity profiles in an elastic tube, at low frequency, $\Omega = 1.0$, as in Fig.5.9.2, but here in a tube of length $l = L/100$, where L is wavelength. In Fig.5.9.2, by comparison, the tube length and wavelength are equal.

in an elastic tube compared with that in a rigid tube. Perform the calculation for three different values of the frequency parameter, namely, $\Omega = 1, 3, 10$, and compare the results visually with those in Fig.5.9.9.

5.11 References and Further Reading

1. Rouse H, Ince S, 1957. History of Hydraulics. Dover Publications, New York.
2. McDonald DA, 1974. Blood Flow in Arteries. Edward Arnold, London.
3. Caro CG, Pedley TJ, Schroter RC, Seed WA, 1978. The Mechanics of the Circulation. Oxford University Press, Oxford.
4. Milnor WR, 1989. Hemodynamics. Williams and Wilkins, Baltimore.
5. Lighthill M, 1975. Mathematical Biofluiddynamics. Society for Industrial and Applied Mathematics, Philadelphia.
6. Fung YC, 1984. Biodynamics: Circulation. Springer-Verlag, New York.
7. McLachlan NW, 1955. Bessel Functions for Engineers. Clarendon Press, Oxford.

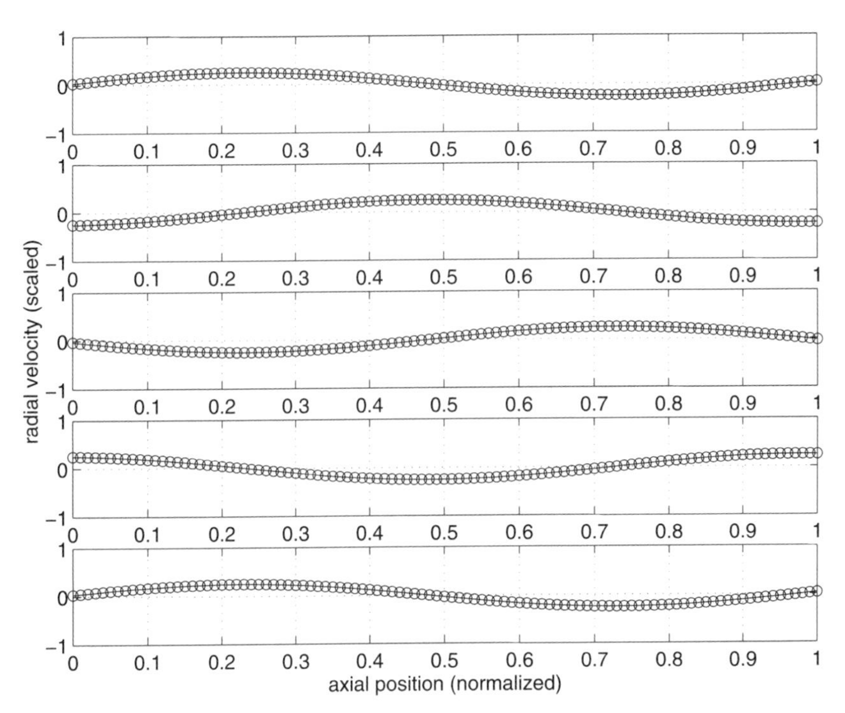

FIGURE 5.9.7. Radial velocity v at the tube wall at low frequency, $\Omega = 1.0$. In each panel the velocity is shown over one wavelength L (normalized to 1.0) along the tube. Different panels represent different times within the oscillatory cycle, with $\omega t = 0$ at the top, then increasing by 90° in each subsequent panel. The velocity is normalized in terms of the maximum velocity in steady flow, $\hat{u}_s$, and further scaled by the factor $2a\omega/i\Lambda^2 c$, so that the quantity shown is only $[1 - Gg]e^{i\omega(t-x/c)}$ (see text). The results mimic the movement of the tube wall within the oscillatory cycle and over one wavelength L along the tube.

8. Watson GN, 1958. A Treatise on the Theory of Bessel Functions. Cambridge University Press. Cambridge.

9. Sechler EE, 1968. Elasticity in Engineering. Dover Publications, New York.

10. Wempner G, 1973. Mechanics of Solids With Applications to Thin Bodies. McGraw-Hill, New York.

11. Shames IH, Cozzarelli FA, 1992. Elastic and Inelastic Stress Analysis. Prentice Hall, Englewood Cliffs, New Jersey.

12. Bradley GL, 1975. A Primer of Linear Algebra. Prentice Hall, Englewood Cliffs, New Jersey.

13. Noble B, Daniel JW, 1977. Applied Linear Algebra. Prentice Hall, Englewood Cliffs, New Jersey.

14. Lay DC, 1994. Linear Algebra and its Applications. Addison-Wesley, Reading, Massachusetts.

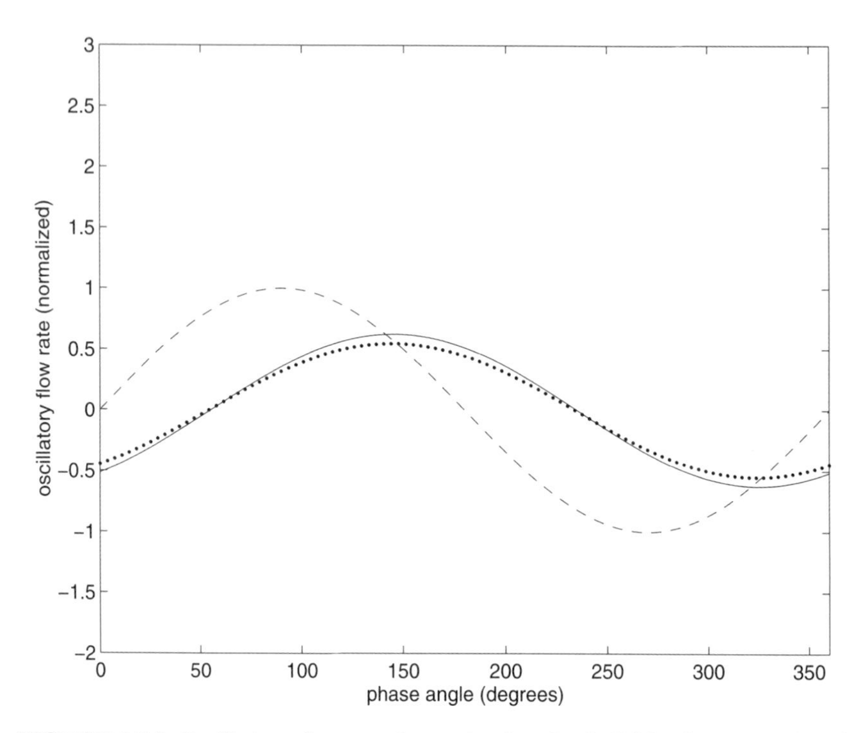

FIGURE 5.9.8. Oscillatory flow rate in an elastic tube (solid line) compared with that in a rigid tube (dotted line), at moderate frequency, $\Omega = 3.0$. The dashed line shows the corresponding pressure gradient. A small difference exists between the two cases at this frequency, with the flow rate in an elastic tube reaching a somewhat higher peak than that in a rigid tube.

15. Morgan GW, Kiely JP, 1954. Wave propagation in a viscous liquid contained in a flexible tube. Journal of Acoustical Society of America 26:323–328.

16. Womersley JR, 1955. Oscillatory motion of a viscous liquid in a thin-walled elastic tube -I: The linear approximation for long waves. Philosophical Magazine 46:199–221.

17. Atabek HB, Lew HS, 1966. Wave propagation through a viscous incompressible fluid contained in an initially elastic tube. Biophysical Journal 6:481–503.

18. Cox RH, 1969. Comparison of linearized wave propagation models for arterial blood flow analysis. Journal of Biomechanics 2:251–265.

19. Ling SC, Atabek HB, 1972. A nonlinear analysis of pulsatile flow in arteries. Journal of Fluid Mechanics 55:493–511.

20. Korteweg DJ, 1878. Über die Fortpflanzungsgeschwindigkeit des Schalles in elastischen Rohren. Annalen der Physik und Chemie 5:525-542.

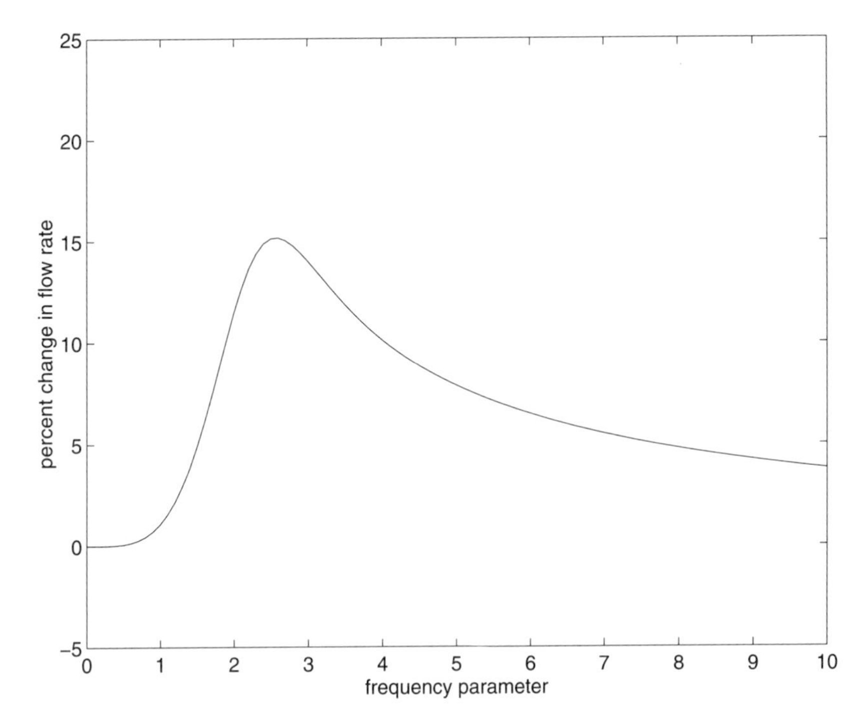

FIGURE 5.9.9. Percentage difference between the peak of flow rate in an elastic tube from that in a rigid tube. Peak flow rate is higher in the elastic tube, at all frequencies. Wall movements in an elastic tube make it somewhat "easier" for the flow to move through the tube.

21. Lamb H, 1897. On the velocity of sound in a tube, as affected by the elasticity of the walls. Memoirs and Proceedings, Manchester Literary and Philosophical Society A42:1-16.

22. Witzig K, 1914. Über erzwungene Wellenbewegungen zäher, inkompressibler Flüssigkeiten in elastischen Rohren. Inaugural Dissertation, Universität Bern, K.J. Wyss, Bern.

23. Lambossy P, 1950. Apercu historique et critique sur le probleme de la propagation des ondes dans un liquide compressible enferme dans un tube elastique. Helvetica Physiologica et Pharmalogica Acta 8:209–227.

6

Wave Reflections

6.1 Introduction

Solutions of the equations for pulsatile flow in an *elastic* tube considered in Chapter 5 produce a flow field that differs only slightly from the corresponding solutions in a *rigid* tube. The elastic tube solution is based on a number of simplifying assumptions, in particular, that the tube wall is thin and that it is only slightly elastic, so that radial displacements of the wall are small and wave speed is high. These assumptions are reasonably well satisfied in the cardiovascular system. Another assumption *implied* in the elastic tube solution is that the tube wall is not "tethered," meaning that it is free to move under the forces of the flow field. In the cardiovascular system many vessels are in fact tethered to surrounding tissue to various degrees, but the effect of this is mostly to add to the mass and stiffness of the vessel wall [1–3]. The first of these may contradict somewhat the assumption of thin tube wall, the second may support the assumption of small elasticity.

Thus on the whole the results of the previous two chapters may suggest that flow in a rigid tube is a reasonable model for flow in an elastic tube, because the velocity fields obtained in the two cases differ only slightly from each other and because the assumptions on which the model is based can be reasonably justified. This conclusion would be in serious error, however, because pulsatile flow in an elastic tube differs in a fundamental way from that in a rigid tube. In an elastic tube, no matter how small the elasticity of the tube, flow propagates down the tube in the form of a *wave*, and in

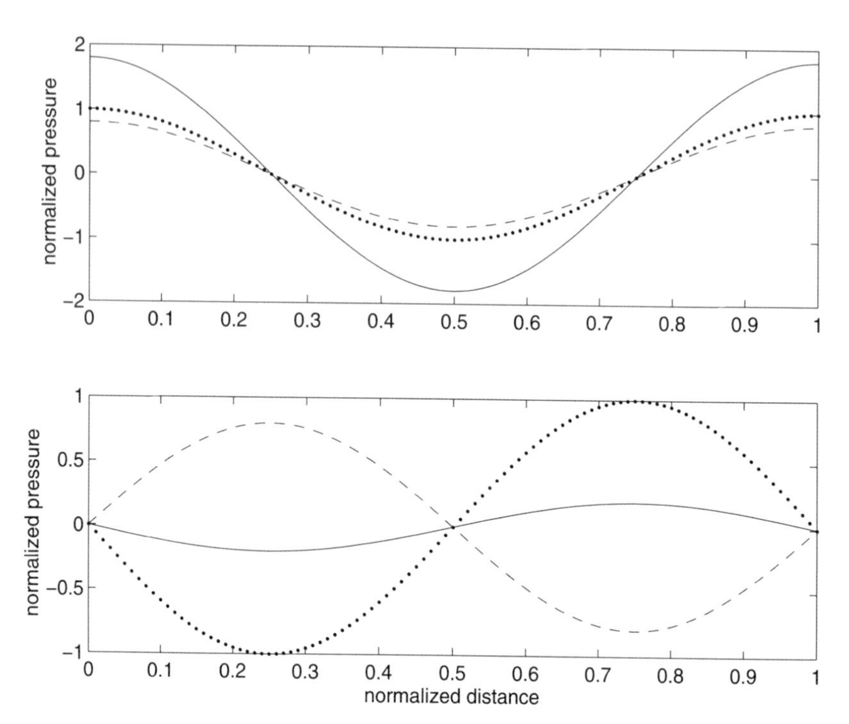

FIGURE 6.1.1. Effect of wave reflection in a tube. Pressure distribution is shown in terms of normalized distance $\overline{x}$ along the tube, where $\overline{x} = 0$ at entrance and $\overline{x} = 1.0$ at exit. A forward pressure wave (dotted line) is reflected at exit, producing a backward moving wave (dashed line). The pressure distribution in the tube is the *sum* of the two waves (solid line), which is thus greatly affected by the nature and extent of the reflected wave. In the upper panel a forward cosine wave is shown reflected at 80%. In the lower panel the same is shown for a sine wave. The length of the wave in both cases is exactly equal to the tube length. We see later how all of these factors affect critically the final pressure distribution in the tube.

the face of any obstacles this wave can be *reflected*. Wave reflections are omnipresent in pulsatile blood flow because all blood vessels have some elasticity and because the vascular tree is dominated by obstacles in the form of vascular junctions. Elasticity of the vessels sustains the wave motion and the ubiquitous vascular junctions constantly produce wave reflections throughout the system [4–7].

In a rigid tube the possibility of wave reflections does not exist in the sense that there is no wave motion in this case. In a rigid tube the wave speed c as defined by the Moen–Korteweg formula (Eq.5.1.1) becomes infinite because Young's modulus E is infinite for a rigid material. Wave propagation at infinite speed is equivalent to changes in flow or pressure being transmitted instantly to every part of the tube. Thus "propagation" in this limit degenerates into "bulk motion" whereby the entire body of

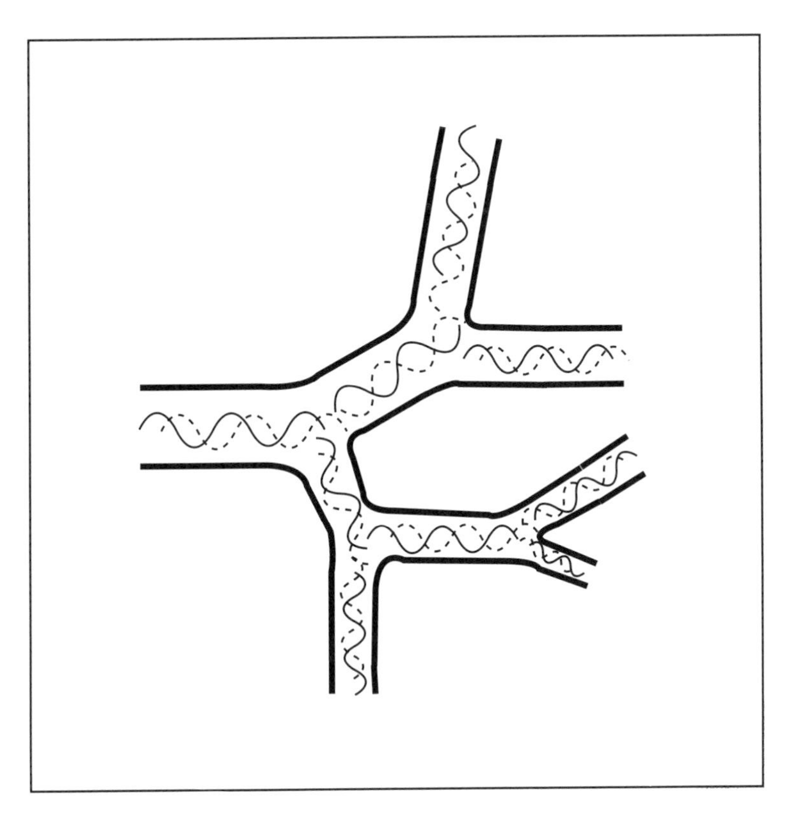

FIGURE 6.1.2. In a branching vascular structure each junction acts as a reflection site, resulting in a bewildering array of forward and backward moving waves. Analysis of the pressure distribution along the tree is possible only if the tube segments comprising the tree are treated as one–dimensional "transmission lines."

fluid is moving in unison (Fig.4.6.1). In the face of obstacles such bulk motion is "disrupted" in some transient way, rather than "reflected" in the same way that a wave is reflected, because in this case there is no wave.

Wave reflections in an elastic tube have the effect of modifying the pressure and flow within the tube because the reflected waves combine with the forward waves (Fig.6.1.1) to produce a new pressure–flow relationship [8]. If there are many reflected waves from many different reflection sites, this relationship may not be easy to predict or compute. Thus the results of the previous two chapters do not depict the most important difference between pulsatile flow in rigid and elastic tubes, because the analysis in both chapters does not include the effects of wave reflections.

In order to take the effects of wave reflections into account, the method of analysis must be simplified because the problems of interest typically involve the superposition of many pressure and flow waves. The method of the previous chapter describes the propagation of a *single* pressure or flow wave in full details and in the entire cross section of the tube. Using the same method for a large number of waves in a large number of tubes as in a vascular tree structure would make the problem intractable (Fig.6.1.2). In fact the method produces far more information than would normally be required to determine the main effects of wave reflections.

The most important effects of wave reflections in a tube are manifest in terms of changes in pressure and flow *in the axial direction*; thus, full details of the flow in cross sections of the tube are not required. Instead, flow properties can be *averaged* over cross sections of the tube to become functions of only one space variable x instead of x and r. A method of analysis that follows this avenue successfully is indeed a one–dimensional method based on the so called transmission line theory [9,10]. We derive and use this method in the present chapter.

6.2 One–Dimensional Wave Equations

The one dimensional method of analysis begins with the equations used for pulsatile flow in an elastic tube, namely, Eqs.5.1.4–6:

$$\rho\frac{\partial u}{\partial t}+\frac{\partial p}{\partial x}=\mu\left(\frac{\partial^2 u}{\partial r^2}+\frac{1}{r}\frac{\partial u}{\partial r}\right)$$

$$\rho\frac{\partial v}{\partial t}+\frac{\partial p}{\partial r}=\mu\left(\frac{\partial^2 v}{\partial r^2}+\frac{1}{r}\frac{\partial v}{\partial r}-\frac{v}{r^2}\right)$$

$$\frac{\partial u}{\partial x}+\frac{\partial v}{\partial r}+\frac{v}{r}=0$$

As they stand these equations are two dimensional in space in the sense that the dependent variables are functions of *two* space variables, namely, x and r. In the present analysis the equations are turned one dimensional by eliminating the dependence on r. This is achieved in essence by integrating over a cross section of the tube so that the main dependent variable is changed from velocity to *flow rate*, and the equation in the radial direction is then no longer required. It is important to note that the latter is not the same as taking the radial velocity v to be identically zero as it is in the rigid tube case. In the present analysis v is not zero even though the radial direction has been eliminated. How this is done we see in details below.

Each term in the first and third equations is multiplied by $2\pi r$ and integrated from $r=0$ to $r=a$, that is,

$$2\pi\rho\int_0^a r\frac{\partial u}{\partial t}dr+2\pi\int_0^a r\frac{\partial p}{\partial x}dr$$

$$=2\pi\mu\int_0^a r\left(\frac{\partial^2 u}{\partial r^2}+\frac{1}{r}\frac{\partial u}{\partial r}\right)dr \qquad (6.2.1)$$

$$2\pi \int_0^a r \frac{\partial u}{\partial x} dr + 2\pi \int_0^a r \left(\frac{\partial v}{\partial r} + \frac{v}{r} \right) dr = 0 \tag{6.2.2}$$

To these governing equations is added the "matching" boundary condition

$$v(a,t) = \frac{\partial a}{\partial t} \tag{6.2.3}$$

which equates the radial velocity of the fluid at the tube wall to the rate of change of the tube radius. This boundary condition is central to the present analysis, because it ensures that although the radial direction is eliminated, the combined effect of radial velocity and elasticity of the tube wall is preserved, and hence the mechanism of wave propagation is preserved. Furthermore, using this boundary condition, the last integral in Eq.6.2.2 becomes

$$\begin{aligned} 2\pi \int_0^a r \left(\frac{\partial v}{\partial r} + \frac{v}{r} \right) dr &= 2\pi \int_{r=0}^{r=a} d(vr) \\ &= 2\pi a v(a) \\ &= \frac{dA}{dt} \end{aligned} \tag{6.2.4}$$

where

$$A(t) = \pi a^2(t) \tag{6.2.5}$$

is cross sectional area of the tube. Using this condition, and noting that

$$2\pi \int_0^a r \frac{\partial u}{\partial t} dr = \frac{\partial q}{\partial t} \tag{6.2.6}$$

$$2\pi \int_0^a r \frac{\partial u}{\partial x} dr = \frac{\partial q}{\partial x} \tag{6.2.7}$$

$$2\pi \frac{\mu}{\rho} \int_0^a r \left(\frac{\partial^2 u}{\partial r^2} + \frac{1}{r}\frac{\partial u}{\partial r} \right) dr = \frac{-2\pi a}{\rho} \tau_w \tag{6.2.8}$$

$$2\pi \int_0^a r \left(\frac{\partial v}{\partial r} + \frac{v}{r} \right) dr = 2\pi a v(a) \tag{6.2.9}$$

where q is flow rate through the tube and τ_w is shear stress exerted by the fluid on the tube wall, that is

$$q = 2\pi \int_0^a r u dr \tag{6.2.10}$$

$$\tau_w = -\mu \left(\frac{\partial u}{\partial r} \right)_{r=a} \tag{6.2.11}$$

Thus Eqs.6.2.1,2 finally become

$$\frac{\partial q}{\partial t} + \frac{A}{\rho}\frac{\partial p}{\partial x} = \frac{-2\pi a}{\rho}\tau_w \tag{6.2.12}$$

$$\frac{\partial q}{\partial x} + \frac{\partial A}{\partial t} = 0 \tag{6.2.13}$$

The basic form of transmission line theory is based on *inviscid* flow where $\tau_w = 0$ and the wave speed c_0 is given by

$$c_0^2 = \frac{A}{\rho}\frac{\partial p}{\partial A} \tag{6.2.14}$$

Noting that

$$\frac{\partial A}{\partial t} = \frac{\partial A}{\partial p}\frac{\partial p}{\partial t} = \frac{A}{\rho c_0^2}\frac{\partial p}{\partial t} \tag{6.2.15}$$

Eqs.6.2.12,13 under these simplifying conditions reduce to

$$\frac{\partial q}{\partial t} + \frac{A}{\rho}\frac{\partial p}{\partial x} = 0 \tag{6.2.16}$$

$$\frac{\partial q}{\partial x} + \frac{A}{\rho c_0^2}\frac{\partial p}{\partial t} = 0 \tag{6.2.17}$$

In many wave propagation studies this basic inviscid form of the equations has the advantage of isolating the effects of wave reflections from the effects of viscosity. The main effects of viscosity on wave propagation, as we saw in the previous chapter, is to reduce the speed and amplitude of the traveling wave. These effects are fairly predictable and can in fact be easily reinstated in the one–dimensional equations [9,11–13], but it is convenient to leave them out while the focus is on wave reflections. With this in mind, cross differentiation of Eqs.6.2.16,17 finally leads to

$$\frac{\partial^2 p}{\partial t^2} = c_0^2\frac{\partial^2 p}{\partial x^2} \tag{6.2.18}$$

$$\frac{\partial^2 q}{\partial t^2} = c_0^2\frac{\partial^2 q}{\partial x^2} \tag{6.2.19}$$

the coefficients A/ρ and $A/\rho c_0^2$ being treated as constants in the differentiation. Pressure and flow are governed by the *same* one–dimensional wave equation and propagate with the same wave speed. This does not mean that they are *in phase*, however, as we see in what follows.

6.3 Basic Solution of Wave Equation

Because the governing equations for pressure and flow are identical wave equations, we consider here only the equation for the pressure. Solution of

Eq.6.2.18 is obtained by separation of variables, that is, by writing

$$p(x,t) = p_x(x)p_t(t) \tag{6.3.1}$$

The form of the driving pressure applied at entrance to the tube must be specified to proceed with the solution. We consider a complex exponential form as in the previous chapter, that is, we take

$$p_a(t) = p_0 e^{i\omega t} = p_x(0)p_t(t) \tag{6.3.2}$$

which clearly implies that

$$p_x(0) = p_0, \quad \text{and} \quad p_t(t) = e^{i\omega t} \tag{6.3.3}$$

where p_0 is the amplitude of the applied oscillatory pressure at the tube entrance. The second result indicates that the time-dependent part of pressure within the tube must have the same functional form as the pressure applied at the tube entrance. Thus the expression for the pressure is now

$$p(x,t) = p_x(x)e^{i\omega t} \tag{6.3.4}$$

and it remains to determine only the x-dependent part of the pressure. Substituting Eq.6.3.4 into Eq.6.2.18 leads to an ordinary differential equation for $p_x(x)$, namely,

$$\frac{d^2p_x}{dx^2} + \frac{\omega^2}{c_0^2}p_x = 0 \tag{6.3.5}$$

This equation is a standard second-order linear differential equation with the general solution [14]

$$p_x(x) = Be^{-i\omega x/c_0} + Ce^{i\omega x/c_0} \tag{6.3.6}$$

where B, C are arbitrary constants. With this the complete expression for the pressure, from Eq.6.3.4, becomes

$$p(x,t) \quad = \quad p_x(x)e^{i\omega t} = Be^{i\omega(t-x/c_0)} + Ce^{i\omega(t+x/c_0)} \tag{6.3.7}$$

The first part of this solution represents a wave traveling in the positive x direction at a speed c_0. This interpretation is gained by observing that the pressure is constant when

$$x = c_0 t \tag{6.3.8}$$

which represents a point moving in the positive x direction at a speed c_0.

The second part of the solution represents a wave traveling with speed c_0 in the *negative* x direction. It is not to be confused with waves returning after being reflected, which we shall deal with fully later. Here the reverse wave arises *at the same time* as the forward wave and the two move symmetrically in opposite directions. Reflected waves, by contrast, consist initially of forward waves arising from the first part of the solution, traveling forward to a point where they meet an obstacle and where they give rise to reflected waves traveling back.

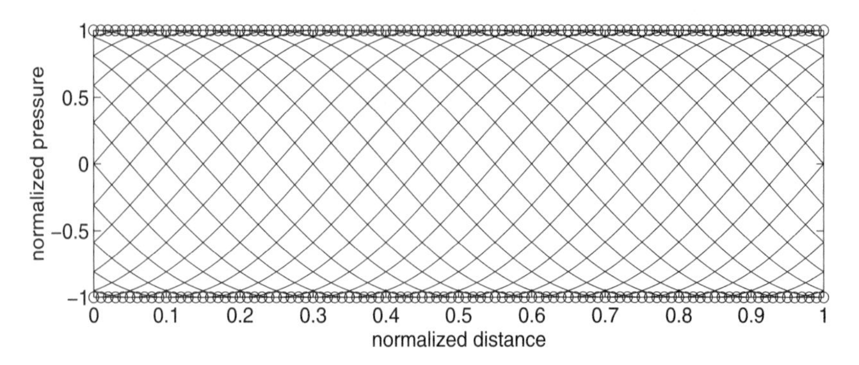

FIGURE 6.3.1. Progression of a sinusoidal pressure wave along an elastic tube in the absence of wave reflections. Individual curves represent the real part of the pressure distribution in the tube at different times within the oscillatory cycle, plotted in terms of normalized distance $\overline{x}$ along the tube, where $\overline{x} = 0$ at entrance and $\overline{x} = 1.0$ at exit. The outer envelope (small circles) represents the limits of that distribution at all times and at different points along the tube. It represents the amplitude of *time* oscillations of pressure at that position along the tube. The ideal pressure amplitude distribution seen in this case, characterized by a straight envelope along the tube, is singular in the sense that it is only possible in the absence of wave reflections.

Pulsatile flow in a tube is typically driven by a pulsating pressure source at the tube entrance. In the context of this analytical solution, this pressure source gives rise to two waves starting *simultaneously* from the entrance and traveling in opposite directions. Therefore only the forward wave is physically relevant and we take, from Eq.6.3.7

$$p(x,t) = Be^{i\omega(t-x/c_0)} \tag{6.3.9}$$

The condition at the tube entrance requires that

$$p(0,t) = Be^{i\omega t} = p_0 e^{i\omega t}, \quad \text{hence} \quad B = p_0 \tag{6.3.10}$$

and the solution is finally

$$p(x,t) = p_0 e^{i\omega(t-x/c_0)} \tag{6.3.11}$$

It is convenient to nondimensionalize in terms of the amplitude of the applied pressure at the tube entrance and introduce the notation

$$\overline{p}(x,t) = \frac{p(x,t)}{p_0}, \quad \overline{p}_x(x) = \frac{p_x(x)}{p_0}, \quad \overline{p}_a(t) = \frac{p_a(t)}{p_0} \tag{6.3.12}$$

so that the solution can be put in the nondimensional form

$$\overline{p}(x,t) = e^{i\omega(t-x/c_0)} \tag{6.3.13}$$
$$= \overline{p}_x(x)e^{i\omega t} \tag{6.3.14}$$

We shall find that the second form of the solution (Eq.6.3.14) is more useful. In this form it is seen that the pressure wave within the tube is composed of two periodic functions, one in space and one in time. At any fixed point in time, the pressure distribution within the tube is described by $\overline{p}_x(x)$. At any fixed position within the tube, the pressure oscillation is described by $e^{i\omega t}$, which is the same oscillatory function in time as the applied pressure at the tube entrance. However, the phase and amplitude of that oscillation depend on $\overline{p}_x(x)$ and therefore this entity plays a central role in the physical description of the propagating wave.

In particular, because the solution $\overline{p}(x,t)$ was obtained in complex form for a complex applied pressure $\overline{p}_a(t)$ at the tube entrance, the real and imaginary parts of $\overline{p}(x,t)$ correspond to the real and imaginary parts of $\overline{p}_a$. Now the real and imaginary parts and the amplitude of $\overline{p}(x,t)$ are given by

$$\begin{aligned} \Re\{\overline{p}(x,t)\} &= \Re\{\overline{p}_x(x)e^{i\omega t}\} \\ \Im\{\overline{p}(x,t)\} &= \Im\{\overline{p}_x(x)e^{i\omega t}\} \\ |\overline{p}(x,t)| &= |\overline{p}_x(x)e^{i\omega t}| = |\overline{p}_x(x)| \end{aligned} \tag{6.3.15}$$

while the real and imaginary parts and the amplitude of $\overline{p}_a(t)$ are given by

$$\Re\{\overline{p}_a\} = \cos\omega t, \quad \Im\{\overline{p}_a\} = \sin\omega t, \quad |\overline{p}_a| = 1.0 \tag{6.3.16}$$

Equations 6.3.15 show the important role that $\overline{p}_x(x)$ plays in determining the characteristics of the propagating pressure wave. The complex form of $\overline{p}_x(x)$ determines the final form of the real and imaginary parts of the pressure within the tube as seen from the first two equations, while the amplitude of $\overline{p}_x(x)$ determines the amplitude of the time oscillations at fixed positions along the tube as is seen from the third equation. Thus the *distribution* of $|\overline{p}_x(x)|$ along the tube is an important measure of pressure oscillations within the tube.

In particular, in the present case where wave reflections are absent, this distribution is given by, from Eqs.6.3.13,14,

$$|\overline{p}_x(x)| = |e^{-i\omega x/c_0}| = 1.0 \tag{6.3.17}$$

which indicates that the time oscillations of the nondimensional pressure within the tube have an amplitude of 1.0 *at every position along the tube*, as illustrated in Fig.6.3.1.

We shall see that this *uniform* distribution only exists in the absence of wave reflections and in the absence of viscosity, and it therefore serves as an important reference state as we proceed to consider wave reflections. In the absence of viscosity, any departure from this state can be attributed directly and entirely to the effects of wave reflections. This advantage provides a good justification for considering the effects of wave reflections in isolation from the effects of viscosity as we do in what follows.

6.4 Primary Wave Reflections in a Tube

When an oscillatory pressure $p_a(t)$, as in Eq.6.3.2, is applied at the entrance of a tube of length l, pressure oscillations will travel toward the other end of the tube in the form of a propagating wave as described in the previous section. In the presence of any change in conditions at that end of the tube, such as a junction with a smaller or larger tube or a division into branches, part of the wave will be reflected back toward the entrance. Depending on conditions at the entrance, part of this backward traveling wave may then itself be reflected back towards the other end of the tube, and the process in principle may continue indefinitely.

In this section we consider only the first of these reflections, which we refer to as "primary" wave reflections. Subsequent reflections shall be referred to as "secondary" wave reflections and will be considered in the next section.

In order to distinguish between forward and backward traveling waves, subscripts f and b are used. Thus the result of the previous section (Eq.6.3.13) can now be used to represent the initial forward traveling wave and written as

$$\overline{p}_f(x,t) = e^{i\omega(t-x/c_0)} \tag{6.4.1}$$

The reflected backward traveling wave will be of the same form but proceeding in the negative x direction, that is,

$$\overline{p}_b(x,t) = Be^{i\omega(t+x/c_0)} \tag{6.4.2}$$

where B is a constant determined by conditions at the reflecting end of the tube. These conditions are usually expressed in terms of a so-called "reflection coefficient" R, which represents the proportion of backward to forward traveling pressure waves at the reflection site ($x = l$), that is,

$$R = \frac{\overline{p}_b(l,t)}{\overline{p}_f(l,t)} \tag{6.4.3}$$

Thus R is a measure of the "extent" or "severity" of the reflection. $R = 1$ represents "total" reflection whereby the backward wave is equal to the forward wave in its entirety, while a small value of R represents a small reflection whereby the backward wave is only a small fraction of the forward wave. Using Eq.6.4.2, the backward wave evaluated at the reflection site is given by

$$\overline{p}_b(l,t) = Be^{i\omega(t+l/c_0)} \tag{6.4.4}$$

while from Eq.6.4.3,1 we have

$$\begin{aligned}\overline{p}_b(l,t) &= R\overline{p}_f(l,t)\\ &= Re^{i\omega(t-l/c_0)}\end{aligned} \tag{6.4.5}$$

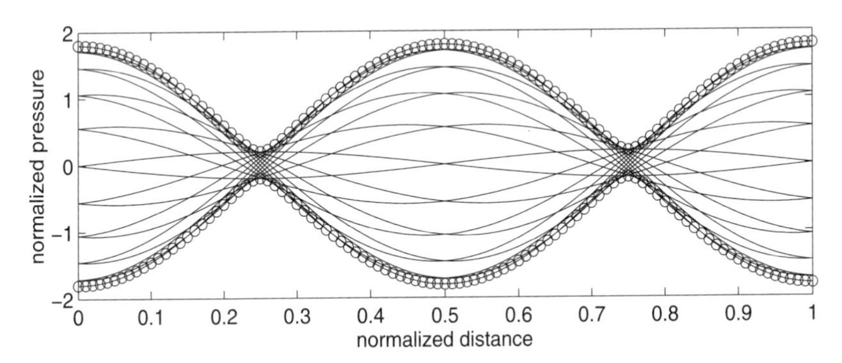

FIGURE 6.4.1. Progression of a sinusoidal pressure wave along an elastic tube in the presence of 80% wave reflections ($R = 0.8$), to be compared with the corresponding results in Fig.6.3.1 for the case in which wave reflections are absent. Individual curves represent the real part of the total (forward plus backward) pressure distribution in the tube at different times within the oscillatory cycle, plotted in terms of normalized distance $\overline{x}$ along the tube, where $\overline{x} = 0$ at entrance and $\overline{x} = 1.0$ at exit. The outer envelope (small circles) represents the limits of that distribution at all times and at different points along the tube. It represents the amplitude of *time* oscillations of pressure at that position along the tube. The shape of the envelope is affected critically by the ratio of wavelength to tube length ($\overline{L}$). In this figure $\overline{L} = 1.0$.

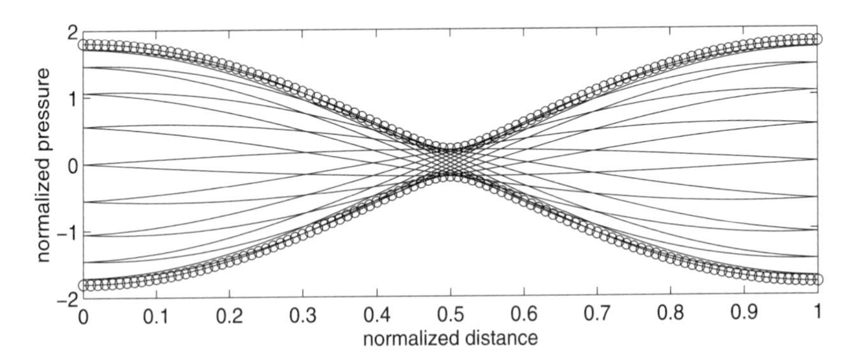

FIGURE 6.4.2. Progression of a sinusoidal pressure wave along an elastic tube in the presence of 80% wave reflections ($R = 0.8$), as in Fig.6.4.1, but here the ratio of wave length to tube length $\overline{L} = 2.0$.

The two results give

$$B = Re^{-2i\omega l/c_0} \qquad (6.4.6)$$

and substituting this value of B in Eq.6.4.2 gives the backward wave in its final form

$$\overline{p}_b(x,t) = Re^{i\omega(t+x/c_0-2l/c_0)} \qquad (6.4.7)$$

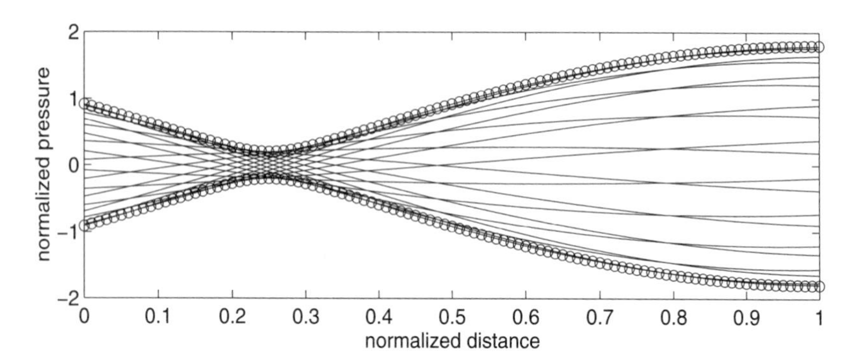

FIGURE 6.4.3. Progression of a sinusoidal pressure wave along an elastic tube in the presence of 80% wave reflections ($R = 0.8$), as in Fig.6.4.1, but here the ratio of wave length to tube length $\overline{L} = 3.0$.

At any point in time and position along the tube the prevailing pressure is the *sum* of the forward and backward traveling waves (Fig.6.1.1), that is,

$$\overline{p}(x,t) = \overline{p}_f(x,t) + \overline{p}_b(x,t) \tag{6.4.8}$$

$$= e^{i\omega(t-x/c_0)} + Re^{i\omega(t+x/c_0-2l/c_0)} \tag{6.4.9}$$

As in the case of no reflections, this can be put in the form

$$\overline{p}(x,t) = \overline{p}_x(x)e^{i\omega t} \tag{6.4.10}$$

where $\overline{p}_x(x)$ has the same important interpretation as before, but now is given by

$$\overline{p}_x(x) = e^{-i\omega x/c_0} + Re^{i\omega(x-2l)/c_0} \tag{6.4.11}$$

It is clear from this result that $|\overline{p}_x(x)|$ is no longer constant along the tube as it is in the absence of wave reflections. In fact it can only become constant if either $R = 0$ or l is infinite. The first corresponds to perfect conditions at the end of the tube so that wave reflections are absent, and the second corresponds to the end of the tube being infinitely far away so that any reflections from that end cannot return in finite time.

The extent to which the pressure distribution in a tube is affected by wave reflections depends also on the frequency ω, as is apparent from Eq.6.4.11. Because frequency is related to the wavelength L by

$$L = \frac{2\pi c_0}{\omega} \tag{6.4.12}$$

the result in Eq.6.4.11 can be put in terms of wave and tube lengths

$$\overline{p}_x(x) = e^{-2\pi ix/L} + Re^{2\pi i(x-2l)/L} \tag{6.4.13}$$

Furthermore, it is convenient to introduce

$$\overline{x} = \frac{x}{l}, \quad \overline{L} = \frac{L}{l} \tag{6.4.14}$$

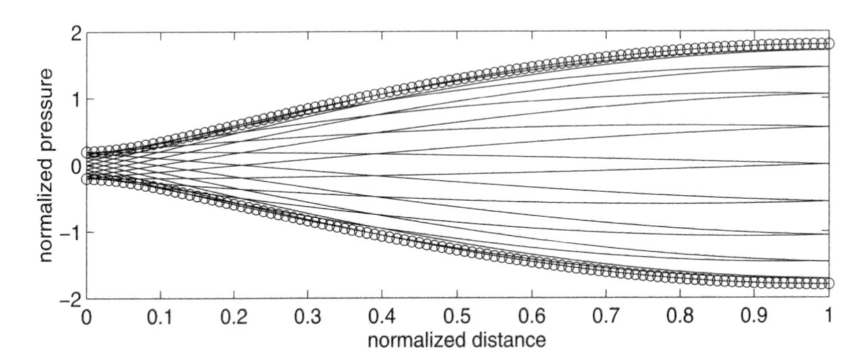

FIGURE 6.4.4. Progression of a sinusoidal pressure wave along an elastic tube in the presence of 80% wave reflections ($R = 0.8$), as in Fig.6.4.1, but here the ratio of wave length to tube length $\overline{L} = 4.0$.

whereby the result takes the normalized form

$$\overline{p}_x(\overline{x}) = e^{-2\pi i\overline{x}/\overline{L}} + R\left(e^{2\pi i\overline{x}/\overline{L}}\right)\left(e^{-4\pi i/\overline{L}}\right) \tag{6.4.15}$$

The advantage of this form is that as the ratio of wavelength to tube length varies, the full range of positions along the tube is always described by $\overline{x} = 0$ to $\overline{x} = 1.0$. An important reference case to consider is that for which the wave and tube lengths are equal, thus $\overline{L} = 1$, $e^{-4\pi i/\overline{L}} = 1$, and the expression reduces to

$$\overline{p}_x(\overline{x}) = e^{-2\pi i\overline{x}} + Re^{2\pi i\overline{x}} \tag{6.4.16}$$

$$= (R+1)\cos 2\pi\overline{x} + i(R-1)\sin 2\pi\overline{x} \tag{6.4.17}$$

The distribution of $|\overline{p}_x(\overline{x})|$ at different positions along the tube is shown in Fig.6.4.1, to be compared with the constant distribution seen in Fig.6.3.1 where wave reflections are absent. Analytically, from Eq.6.4.16 we now have

$$|\overline{p}_x(\overline{x})| = \sqrt{R^2 + 1 + 2R\cos 4\pi\overline{x}} \tag{6.4.18}$$

compared with the result in Eq.6.3.17 for the case in which wave reflections are absent.

In particular, the value of $|\overline{p}_x(\overline{x})|$ is maximum at $\overline{x} = 0,\ 1/2,\ 1$, and minimum at $\overline{x} = 1/4,\ 3/4$. At the maximum points the forward and backward waves *add*, while at the minimum points they *subtract*.

Because $|\overline{p}_x(\overline{x})|$ represents the amplitude of time oscillations at normalized position $\overline{x}$ along the tube, it is seen that these oscillations will have different amplitude at different positions as long as $R \neq 0$. The distribution of $|\overline{p}_x(\overline{x})|$ is relatively easy to describe in the special case of $\overline{L} = 1$ in Fig.6.4.1 but becomes more complicated for other values of $\overline{L}$ and for specific values of the reflection coefficient R where the forward and backward waves combine in more complicated ways. Results for $\overline{L} = 2, 3, 4, 5$ are shown in Figs.6.4.2–5.

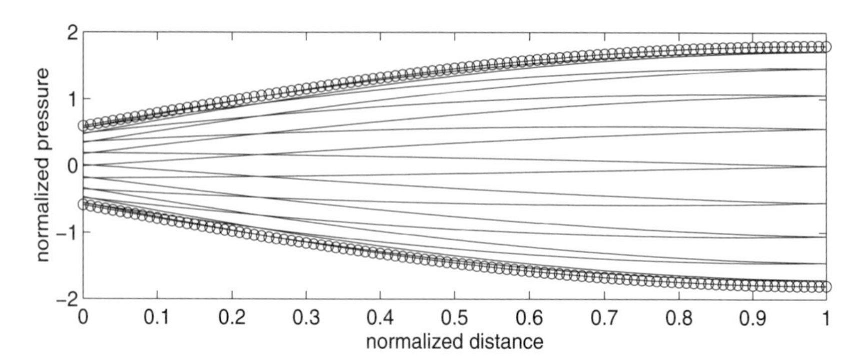

FIGURE 6.4.5. Progression of a sinusoidal pressure wave along an elastic tube in the presence of 80% wave reflections ($R = 0.8$), as in Fig.6.4.1, but here the ratio of wave length to tube length $\overline{L} = 5.0$.

6.5 Secondary Wave Reflections in a Tube

When wave reflections occur at both ends of a tube, the primary backward wave considered in the previous section will be reflected at the tube entrance and part of it will turn to propagate now in the positive x direction toward the other reflecting end of the tube. This process will continue indefinitely, although clearly with diminishing effects since the reflected part of the wave is each time only a fraction of the incident wave [15].

As before, wave propagation within the tube can be represented by two oscillatory functions, one in space and one in time, as in Eq.6.4.10,

$$\overline{p}(x,t) = \overline{p}_x(x)e^{i\omega t}$$

We saw in the previous section that the form of $\overline{p}_x(x)$ begins as a simple complex exponential function when wave reflections are absent, then becomes more composite as the primary reflected wave is added. In this section we pursue this process to its ultimate limit, as reflected waves are now allowed to go back and forth between the two ends of the tube.

To consider this process in detail, let the reflection coefficients at the tube entrance ($x = 0$) and the tube end ($x = l$) be denoted by R_0, R_l, respectively (Fig.6.5.1). We shall also identify the forward and backward waves by subscripts f, b, as before, but now these shall be supplemented by numbers that identify their position in the sequence of repeated reflections. Thus, as in the previous section, we begin with the initial forward wave, from Eq.6.4.1,

$$\overline{p}_{f1}(x,t) = e^{i\omega(t-x)/c_0} \tag{6.5.1}$$

the subscript 1 now being added to indicate that this is the first forward wave, before any reflections have taken place. As before, we write this as the product of two oscillatory functions

$$\overline{p}_{f1}(x,t) = \overline{p}_{xf1}(x)e^{i\omega t} \tag{6.5.2}$$

and because the time oscillatory term does not change with reflections, we shall focus on only the space-dependent term, which for this initial forward wave, from Eq.6.5.1, is given by

$$\overline{p}_{xf1}(x) = e^{-i\omega x/c_0} \tag{6.5.3}$$

Noting that

$$\overline{p}_{xf1}(0) = 1, \quad \overline{p}_{xf1}(l) = e^{-i\omega l/c_0} \tag{6.5.4}$$

the x oscillatory function can be thought of as arising from an input pressure amplitude $p_{xf1}(0)$ at the tube entrance, which is then "operated on" by the complex exponential function representing the translation of that input into a wave form along the tube. We shall find this concept useful as the wave is reflected back and forth, and to make use of it we write Eq.6.5.3 in the form

$$\overline{p}_{xf1}(x) = [\overline{p}_{xf1}(0)]e^{-i\omega x/c_0} \tag{6.5.5}$$

As this wave reaches the other end of the tube ($x = l$), its value at that point, multiplied by the reflection coefficient there, becomes the "input pressure amplitude" for the reflected wave. This wave progresses in the *negative* x direction, and its distance as it moves away from the reflection site is $l - x$ in place of x for the forward wave. The result is

$$\overline{p}_{xb1}(x) = [R_l e^{-i\omega l/c_0}]e^{-i\omega(l-x)/c_0} \tag{6.5.6}$$
$$= R_l e^{-i\omega(2l-x)/c_0} \tag{6.5.7}$$

It will be noted that the sum $\overline{p}_{xf1}(x) + \overline{p}_{xb1}(x)$ is identical with the result obtained in the previous section (Eq.6.4.11) for the pressure wave after only one primary reflection. In this section we continue this process beyond the primary stage.

As the backward wave reaches the entrance of the tube and is allowed to be reflected there, the value of $\overline{p}_{xb1}(x)$ at that point ($x = 0$), multiplied by the reflection coefficient there, becomes the input pressure amplitude for the next forward wave, that is

$$\overline{p}_{xf2}(x) = [R_0 R_l e^{-i\omega 2l/c_0}]e^{-i\omega x/c_0} \tag{6.5.8}$$
$$= R_0 R_l e^{-i\omega(2l+x)/c_0} \tag{6.5.9}$$

The pattern is now established for subsequent reflections. As waves move back and forth between the two ends of the tube, we have

$$\overline{p}_{xb2}(x) = [R_0 R_l^2 e^{-i\omega 3l/c_0}]e^{-i\omega(l-x)/c_0} \tag{6.5.10}$$
$$= R_0 R_l^2 e^{-i\omega(4l-x)/c_0} \tag{6.5.11}$$

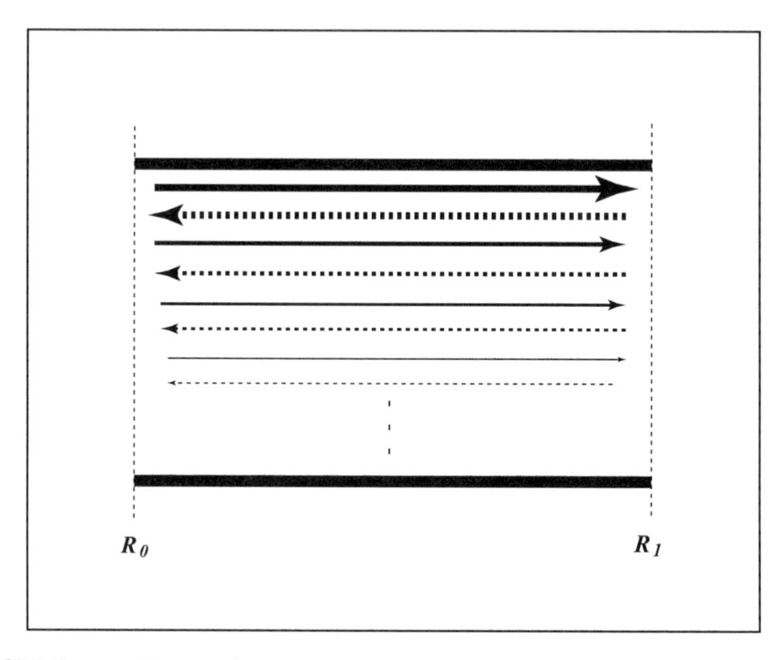

FIGURE 6.5.1. Wave reflections at both ends of a tube. Pressure or flow waves travel back and forth, indefinitely, although each time being reduced in magnitude by an amount determined by the reflection coefficient.

$$\overline{p}_{xf3}(x) = [R_0^2 R_l^2 e^{-i\omega 4l/c_0}]e^{-i\omega x/c_0} \tag{6.5.12}$$
$$= R_0^2 R_l^2 e^{-i\omega(4l+x)/c_0} \tag{6.5.13}$$

$$\overline{p}_{xb3}(x) = [R_0^2 R_l^3 e^{-i\omega 5l/c_0}]e^{-i\omega(l-x/c_0)} \tag{6.5.14}$$
$$= R_0^2 R_l^3 e^{-i\omega(6l-x)/c_0} \tag{6.5.15}$$

The net pressure distribution within the tube consists of the sum of all the forward and backward waves, that is,

$$\begin{aligned}\overline{p}_x(x) &= \overline{p}_{xf1}(x) + \overline{p}_{xf2}(x) + \overline{p}_{xf3}(x) + ... \\ &+ \overline{p}_{xb1}(x) + \overline{p}_{xb2}(x) + \overline{p}_{xb3}(x) + ...\end{aligned} \tag{6.5.16}$$

$$\begin{aligned}&= e^{-i\omega x/c_0} + R_0 R_l e^{-i\omega(2l+x)/c_0} \\ &+ R_0^2 R_l^2 e^{-i\omega(4l+x)/c_0} + ... \\ &+ R_l e^{-i\omega(2l-x)/c_0} + R_0 R_l^2 e^{-i\omega(4l-x)/c_0} \\ &+ R_0^2 R_l^3 e^{-i\omega(6l-x)/c_0} + ...\end{aligned} \tag{6.5.17}$$

$$\begin{aligned}&= e^{-i\omega x/c_0}\{1 + (R_0 R_l e^{-i\omega 2l/c_0}) \\ &+ (R_0 R_l e^{-i\omega 2l/c_0})^2 + (R_0 R_l e^{-i\omega 2l/c_0})^3 + ...\} \\ &+ R_l e^{-i\omega(2l-x)/c_0}\{1 + (R_0 R_l e^{-i\omega 2l/c_0})\end{aligned}$$

$$+ \quad (R_0 R_l e^{-i\omega 2l/c_0})^2 + (R_0 R_l e^{-i\omega 2l/c_0})^3 + ...\} \qquad (6.5.18)$$

The series inside the curly brackets in Eq.6.5.18 is an infinite geometric series in ϵ, where

$$\epsilon = R_0 R_l e^{-i\omega 2l/c_0} < 1 \qquad (6.5.19)$$

and therefore [16]

$$1 + \epsilon + \epsilon^2 + \epsilon^3 + ... = \frac{1}{1-\epsilon} \qquad (6.5.20)$$

Thus Eq.6.5.18 for the pressure finally condenses to

$$\overline{p}_x(x) = \frac{e^{-i\omega x/c_0} + R_l e^{-i\omega(2l-x)/c_0}}{1 - R_0 R_l e^{-i\omega 2l/c_0}} \qquad (6.5.21)$$

Comparing this with the result obtained in the previous section (Eq.6.4.11), it is seen that the two become identical when $R_0 = 0$, that is when there are no reflections from the tube entrance as was assumed in the previous section. Despite this difference, in many studies of wave reflections it is found adequate to use the results of primary reflections only [4,5,17,18]. One justification for this is that the term in the denominator of Eq.6.5.21 contains the product of two reflection coefficients that usually have values, in magnitude, less than 1.0. Furthermore, while the problem of secondary reflections is fairly simple to formulate in a *single tube*, it becomes almost intractable in a complex vascular tree structure consisting of a large number of tube segments and an equally large number of reflection sites. In such cases it is found that analysis of only primary wave reflections is tractable, and this has generally been taken as a reasonable assessment of the way in which wave reflections will affect the flow.

6.6 Pressure–Flow Relations

The ultimate purpose of determining the pressure distribution in a tube is to obtain a measure of the flow within the tube, using some relation between pressure and flow. In *steady* flow, the relation between pressure and flow is dominated and fully determined by viscous resistance. Equations 3.4.3 illustrates the simple relation that exists in this case between the flow rate q_s and the constant pressure gradient k_s. In *pulsatile* flow through a *rigid* tube the relation between pressure and flow is again affected by viscous resistance but now also by inertia of the fluid because of repeated acceleration and deceleration. Thus the *frequency* of pulsation becomes an added factor in the relation between pressure and flow, as can be seen in Eq.4.7.5 for the oscillatory flow rate $q_\phi(t)$ in pulsatile flow through a rigid tube.

In pulsatile flow through an *elastic* tube the elastic properties of the tube become yet another added factor in the relation between pressure and flow. Equations 5.9.9 for the flow rate $q(x,t)$ in this case involves not only viscosity of the fluid and frequency of oscillation but also elastic properties of the tube wall, embedded in the elasticity factor G and the wave speed c. Furthermore, the pulsating pressure and flow in this case propagate in the form of progressive waves, as is apparent from the presence of x in Eq.5.9.9, thus admitting the possibility of wave reflections. Results of the previous section demonstrate clearly that wave reflections can affect the pressure distribution in a tube profoundly because of the superposition of forward and backward waves.

This new factor in the relation between pressure and flow is important not only because it can produce major changes in the pressure distribution but also because these changes are not as easily predictable as they are in the case of viscosity or frequency. For this reason, in what follows we focus on this factor in particular and deal with it in *isolation* so as not to mask it by the effects of viscosity and frequency, which have already been examined. This approach has the advantage of making the effects of wave reflections more "visible." As seen in the previous section, by appropriate scaling the changes in pressure distribution can be expressed as deviations from a reference state in which the normalized pressure amplitude has the constant value 1.0 all along the tube. This reference state is attained only in the absence of wave reflections, thus any deviation from it can be identified immediately as resulting from wave reflections.

A basic solution for the flow wave follows much the same lines as that for the pressure. Starting with the solution in Eq.6.3.11 as an initial forward pressure wave, before any reflections

$$p_f(x,t) = p_0 e^{i\omega(t-x/c_0)} \tag{6.6.1}$$

and because pressure and flow are governed by the same wave equations (Eqs.6.2.18,19), we postulate a solution for the corresponding forward flow wave in the form

$$q_f(x,t) = B e^{i\omega(t-x/c_0)} \tag{6.6.2}$$

where B is a constant.

The relation between pressure and flow is governed by Eqs.6.2.16,17. Applying these to the above pressure and flow waves we obtain

$$\frac{\partial q_f}{\partial t} + \frac{A}{\rho}\frac{\partial p_f}{\partial x} = 0 \tag{6.6.3}$$

$$\frac{\partial q_f}{\partial x} + \frac{A}{\rho c_0^2}\frac{\partial p_f}{\partial t} = 0 \tag{6.6.4}$$

Substituting for p_f and q_f, both equations yield the same results, namely,

$$B = \left(\frac{A}{\rho c_0}\right) p_0 \tag{6.6.5}$$

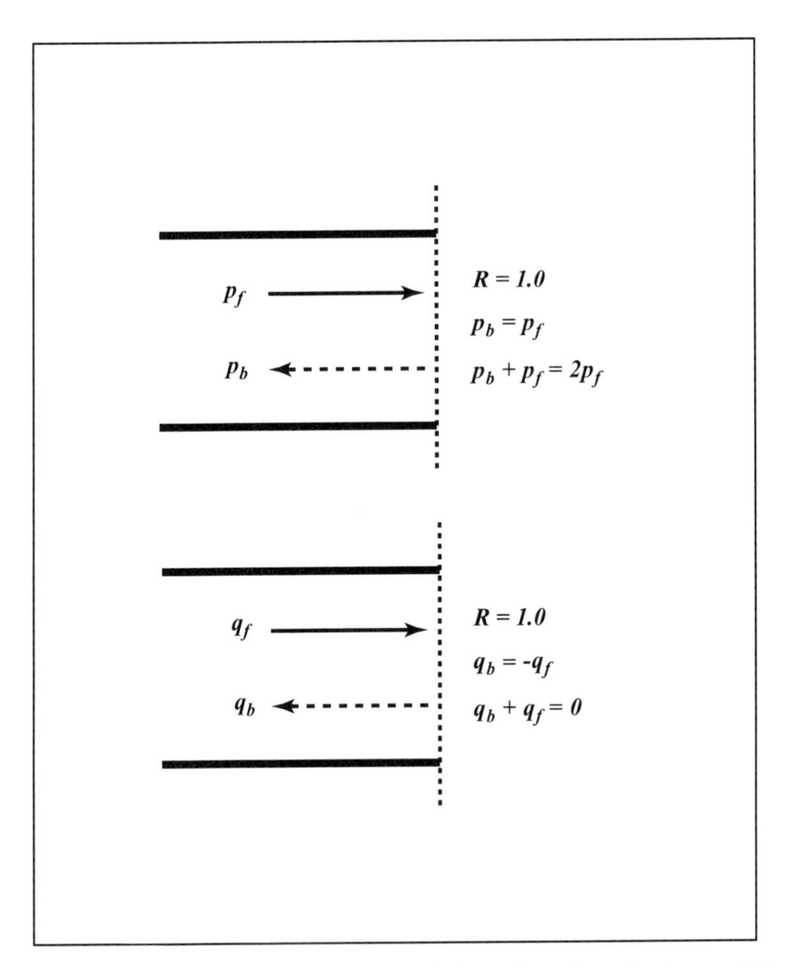

FIGURE 6.6.1. At a completely closed end of a tube, the reflection coefficient $R = 1.0$, the pressure wave is duplicated, and the flow wave is "annihilated."

and

$$q_f(x,t) = \left(\frac{A}{\rho c_0}\right) p_0 e^{i\omega(t-x/c_0)} \tag{6.6.6}$$

$$= \left(\frac{A}{\rho c_0}\right) p_f(x,t) \tag{6.6.7}$$

Similarly, from the result in Eq.6.4.7 for the reflected wave

$$p_b(x,t) = Rp_0 e^{i\omega(t+x/c_0-2l/c_0)} \tag{6.6.8}$$

we postulate a corresponding flow wave

$$q_b(x,t) = Ce^{i\omega(t+x/c_0-2l/c_0)} \tag{6.6.9}$$

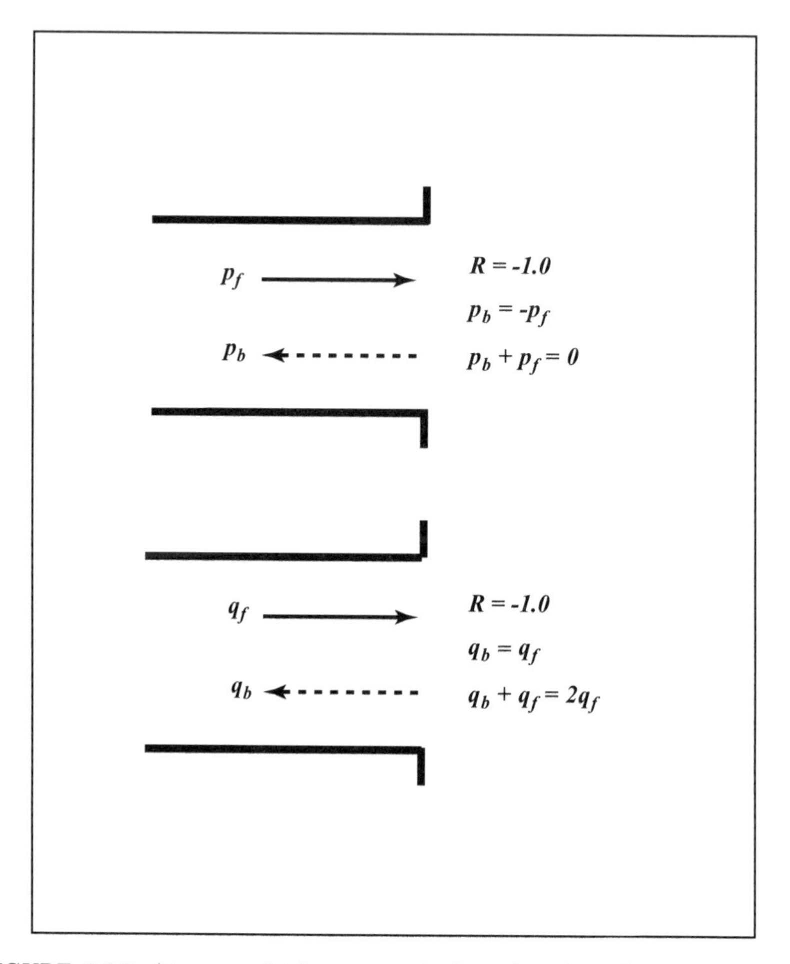

FIGURE 6.6.2. At a completely open end of a tube, the reflection coefficient $R = -1.0$, the pressure wave is "annihilated," and the flow wave is duplicated.

where C is a constant. Applying the governing equations to these waves we obtain

$$\frac{\partial q_b}{\partial t} + \frac{A}{\rho}\frac{\partial p_b}{\partial x} = 0 \tag{6.6.10}$$

$$\frac{\partial q_b}{\partial x} + \frac{A}{\rho c_0^2}\frac{\partial p_b}{\partial t} = 0 \tag{6.6.11}$$

Substituting for p_b and q_b, both equations yield the same results, namely,

$$C = \left(\frac{-A}{\rho c_0}\right) R p_0 \tag{6.6.12}$$

and

$$q_b(x,t) = \left(\frac{-A}{\rho c_0}\right) R p_0 e^{i\omega(t+x/c_0-2l/c_0)} \tag{6.6.13}$$

$$= \left(\frac{-A}{\rho c_0}\right) p_b(x,t) \tag{6.6.14}$$

The quantity

$$Y_0 = \frac{A}{\rho c_0} \tag{6.6.15}$$

is known as the "characteristic admittance" of the tube, and it is seen to be a key parameter in the relation between pressure and flow. Its reciprocal

$$Z_0 = \frac{\rho c_0}{A} \tag{6.6.16}$$

is known as the "characteristic impedance" [10,17,18]. From Eqs.6.6.7,14–16 we then find

$$Y_0 = \frac{q_f(x,t)}{p_f(x,t)} = \frac{-q_b(x,t)}{p_b(x,t)} \tag{6.6.17}$$

$$Z_0 = \frac{p_f(x,t)}{q_f(x,t)} = \frac{p_b(x,t)}{-q_b(x,t)} \tag{6.6.18}$$

from which we draw the interpretation that Y_0 is indeed a measure of the extent to which the tube "admits" flow, while Z_0 is a measure of the extent to which it "impedes" the flow.

The minus sign associated with the backward flow wave indicates that q_b and q_f have opposite signs because of their opposite directions. This issue does not arise in the case of p_b and p_f, because pressure is a *scalar* quantity. An important consequence of this is that when the reflected waves are included in the complete expressions for the pressure and flow waves, we obtain, using Eqs.6.6.1,6,8,13,

$$\begin{aligned} p(x,t) &= p_f(x,t) + p_b(x,t) \\ &= p_0 e^{i\omega(t-x/c_0)} + R p_0 e^{i\omega(t+x/c_0-2l/c_0)} \end{aligned} \tag{6.6.19}$$

$$\begin{aligned} q(x,t) &= q_f(x,t) + q_b(x,t) \\ &= Y_0 \left(p_0 e^{i\omega(t-x/c_0)} - R p_0 e^{i\omega(t+x/c_0-2l/c_0)}\right) \\ &= q_0 e^{i\omega(t-x/c_0)} - R q_0 e^{i\omega(t+x/c_0-2l/c_0)} \end{aligned} \tag{6.6.20}$$

where

$$q_0 = Y_0 p_0 \tag{6.6.21}$$

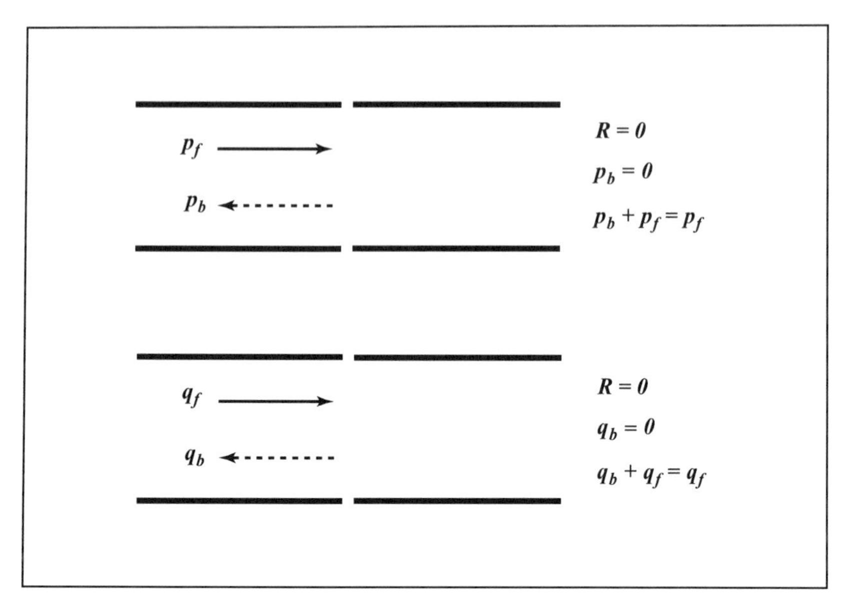

FIGURE 6.6.3. At a completely matched end of a tube, the reflection coefficient $R = 0$, no reflections arise, and the pressure and flow waves are unchanged.

From these results we note that the forward and backward pressure waves *add* while the corresponding flow waves *subtract*. Furthermore, from the definition of the reflection coefficient (Eq.6.4.3) we have

$$R = \frac{p_b(l,t)}{p_f(l,t)} \tag{6.6.22}$$

and in terms of flow waves, using Eq.6.6.17,

$$R = \frac{-q_b(l,t)}{q_f(l,t)} \tag{6.6.23}$$

Thus a value of $R = 1.0$ represents a situation in which the flow wave is *annihilated*, $q_b(l,t) = -q_f(l,t)$, while the pressure wave is *duplicated*, $p_b(l,t) = p_f(l,t)$. This situation would occur when the tube end is completely closed, and the results are therefore consistent with what would be expected on physical grounds (Fig.6.6.1).

A value of $R = -1.0$, similarly, represents a situation in which the flow wave is *duplicated*, $q_b(l,t) = q_f(l,t)$, while the pressure wave is *annihilated*, $p_b(l,t) = -p_f(l,t)$. This situation would occur when the tube end is completely open, and the results are again consistent with what would be expected on physical grounds (Fig.6.6.2).

A value of $R = 0$, finally, represents a situation in which both the flow and the pressure waves are unchanged, that is, $q_b(l,t) = 0$, $p_b(l,t) = 0$. This situation would occur when the tube end is completely matched with another tube of the same properties. In this case no reflections arise, as would be expected on physical grounds (Fig.6.6.3).

6.7 Effective Admittance

Results of the previous section make it clear that in the presence of wave reflections the pressure and flow waves within a tube no longer have the same form (Eqs.6.6.19,20). One of the most important consequences of this is a change in the ratio of flow to pressure and hence a change in the admittance of the tube. The characteristic admittance Y_0 is no longer a measure of the extent to which the tube admits flow, a new "effective admittance" Y_e takes its place.

To determine the effective admittance consider first a single tube extending from $x = 0$ to $x = l$ in which the reflection coefficient at $x = l$ is R. By common convention the effective admittance of the tube is defined as the ratio of flow to pressure at the tube entrance ($x = 0$), so that using Eqs.6.6.19,20 we have

$$Y_e = \frac{q(l,t)}{p(l,t)} = Y_0 \left(\frac{1 - Re^{-2i\omega l/c_0}}{1 + Re^{-2i\omega l/c_0}} \right) \tag{6.7.1}$$

It is clear from this that the effective admittance Y_e will be different from the characteristic admittance Y_0 as long as $R \neq 0$, that is, as long as wave reflections are present.

The reflection coefficient R at $x = l$ may generally result from a transition at $x = l$ from one tube in which the admittance is Y_0 to another in which the admittance is Y_t (Fig.6.7.1). The admittance Y_0 in the first tube is the *characteristic* admittance, while Y_t in the second tube may be the characteristic admittance if wave reflections are absent *in that tube* or the effective admittance if they are present. The reflection coefficient at the junction between the two tubes can be expressed in terms of the difference between the two admittances by applying two conditions at that junction.

The first condition requires that the sum of forward and backward pressure waves evaluated at $x = l$ in the first tube be equal to the forward pressure wave evaluated at $x = 0$ in the second tube, that is,

$$p_{0f}(l,t) + p_{0b}(l,t) = p_{tf}(0,t) \tag{6.7.2}$$

The second condition requires that the *vector* sum of the forward and backward flow at $x = l$ in the first tube be equal to the forward flow at $x = 0$ in the second tube, that is,

$$q_{0f}(l,t) + q_{0b}(l,t) = q_{tf}(0,t) \tag{6.7.3}$$

noting, however, that q_{0f} and q_{0b} have different signs, hence the sum on the left-hand side actually represents the *difference* between the two flows.

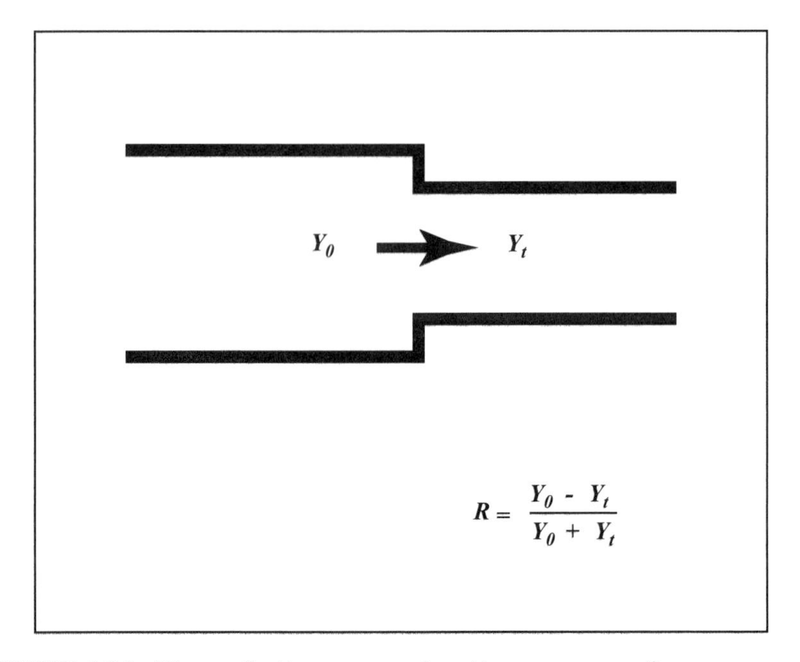

FIGURE 6.7.1. Wave reflections occur when the pressure or flow wave meets a change of admittance from Y_0 in one tube to Y_t in another. The reflection coefficient R is equal to 1.0 when Y_t is zero, -1.0 when Y_t is infinite, and 0 when $Y_t = Y_0$. The three situations are the same as those illustrated in Figs.6.6.1–3.

Substituting from Eqs.6.6.17 for the flow rates into Eq.6.7.3, we then have

$$Y_0 p_{0f}(l,t) - Y_0 p_{0b}(l,t) = Y_t p_{tf}(0,t) \tag{6.7.4}$$

and using Eq.6.7.2 this becomes

$$Y_0 p_{0f}(l,t) - Y_0 p_{0b}(l,t) = Y_t \left[p_{0f}(l,t) + p_{0b}(l,t)\right] \tag{6.7.5}$$

By definition of the reflection coefficient (Eq.6.6.22) we have

$$\frac{p_{ob}(l,t)}{p_{0f}(l,t)} = R \tag{6.7.6}$$

and substitution into Eq.6.7.5 yields

$$R = \frac{Y_0 - Y_t}{Y_0 + Y_t} \tag{6.7.7}$$

We see that the reflection coefficient is zero when $Y_0 = Y_t$, that is, when there is no change of admittance at the junction between the two tubes. The coefficient is equal to 1.0 when $Y_t = 0$, that is, when the admittance of the second tube is zero, there is complete blockage and hence total reflection. The reflection coefficient is equal to -1.0, finally, when Y_t is infinite, that

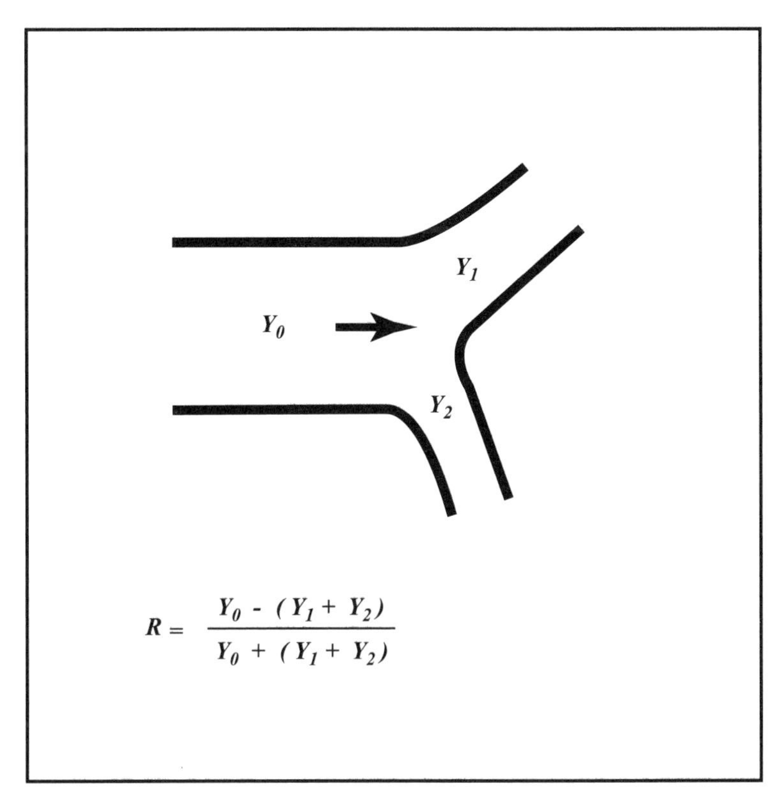

FIGURE 6.7.2. At an arterial bifurcation the pressure or flow wave meets a change of admittance from Y_0 in the parent tube to $Y_1 + Y_2$ in the branch tubes. Reflection at the junction depends on the extent to which $Y_1 + Y_2$ differs from Y_0 (see text).

is, when the second tube is completely open and offers no impedance to flow (or infinite admittance). These three situations correspond with those illustrated in Figs.6.6.1–3.

With the result in Eq.6.7.7 it is possible to eliminate R from the relation between effective and characteristic admittances by writing Eq.6.7.1 in the form

$$Y_e = Y_0 \left\{ \frac{e^{i\theta} - Re^{-i\theta}}{e^{i\theta} + e^{-i\theta}} \right\} \tag{6.7.8}$$

$$= Y_0 \left\{ \frac{(1-R)\cos\theta + i(1+R)\sin\theta}{(1+R)\cos\theta + i(1-R)\sin\theta} \right\} \tag{6.7.9}$$

where

$$\theta = \frac{\omega l}{c_0} \tag{6.7.10}$$

Then noting from Eqs.6.7.7 that

$$1+R=\frac{2Y_0}{Y_0+Y_t}, \quad 1-R=\frac{2Y_t}{Y_0+Y_t} \tag{6.7.11}$$

this finally yields

$$Y_e=Y_0\left\{\frac{Y_t+iY_0\tan\theta}{Y_0+iY_t\tan\theta}\right\} \tag{6.7.12}$$

The change of admittance at a reflection site may result from a "bifurcation," that is, a division of the first tube into two branch tubes, as illustrated in Fig.6.7.2. This situation is particularly important because arterial bifurcations are the main structural units of an arterial tree. If the characteristic admittance of the parent tube is Y_0 and the effective admittances of the two branches are Y_1, Y_2, the results of the previous example are easily extended by noting that in this case

$$Y_t=Y_1+Y_2 \tag{6.7.13}$$

thus Eq.6.7.7 for the reflection coefficient become

$$R=\frac{Y_0-(Y_1+Y_2)}{Y_0+(Y_1+Y_2)} \tag{6.7.14}$$

and Eq.6.7.12 for the effective admittance becomes

$$Y_e=Y_0\left\{\frac{(Y_1+Y_2)+iY_0\tan\theta}{Y_0+i(Y_1+Y_2)\tan\theta}\right\} \tag{6.7.15}$$

These results for a single bifurcation (Eqs.6.7.14,15) provide the foundation for an extension of the results to a vascular tree structure in which an arterial bifuction is the basic structural unit.

6.8 Pressure Distribution in a Vascular Tree Structure

Vascular trees typically consist of repeated bifurcations whereby a vessel segment divides into two branch segments, then each of the branches divides into two branches, and so on. Results obtained at the end of the last section for a single bifurcation (Eqs.6.7.14,15) provide the required basis for extending the analysis to a tree structure containing a large number of bifurcations. However, some issues must be dealt with at first regarding the details of that extension.

The first issue is that of mapping a large number of tube segments that make up a tree structure, in a manner that preserves the hierarchic position of each segment on the tree structure. A simple but effective way of dealing with this is to use a coordinate pair $[j, k]$ to identify the position of each tube segment on the tree structure, where j denotes the "level" or "generation"

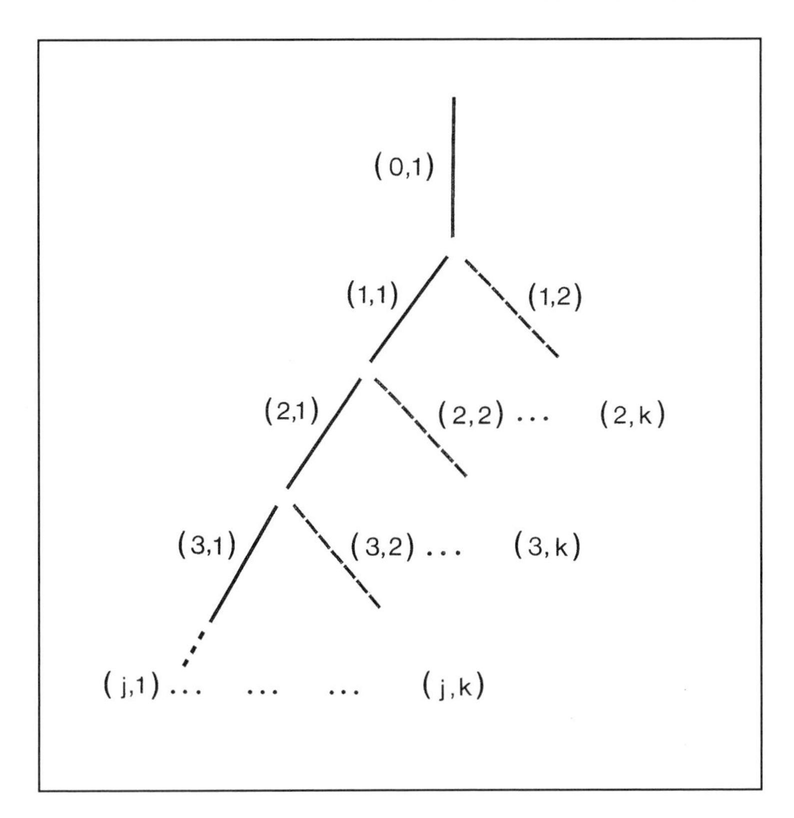

FIGURE 6.8.1. In an open tree structure a coordinate pair (j, k) can be used to identify the position of a tube segment within that structure, where j denotes the level or generation of the tree at which the segment is located and k denotes a sequential position of the segment at that level. The root segment is here denoted by (0,1). In the discussion below it is denoted by [0,0]. From [21].

at which the segment is located and k denotes a sequential position of the segment within that level. Thus if the position of the root segment from which the entire tree derives is denoted by $[0, 0]$, then the position of the two branch segments arising from this segment are $[1, 1]$ and $[1, 2]$, and the positions of their own branches are $[2, 1], [2, 2], [2, 3], [2, 4]$, and so on, as illustrated in Fig.6.8.1.

For a general bifurcation unit within the tree structure, if the position of the parent tube segment is denoted by $[j, k]$, the position of its two branch segments are $[j + 1, 2k - 1]$ and $[j + 1, 2k]$. Using this notation in Eqs.6.7.14,15 for the reflection coefficient and effective admittance, the two results become associated with a bifurcation within a tree structure and take the form

$$R[j, k] = \frac{Y_0[j, k] - (Y_e[j + 1, 2k - 1] + Y_e[j + 1, 2k])}{Y_0[j, k] + (Y_e[j + 1, 2k - 1] + Y_e[j + 1, 2k])} \tag{6.8.1}$$

$$Y_e[j,k] = Y_0[j,k]$$
$$\times \left\{ \frac{(Y_e[j+1,2k-1] + Y_e[j+1,2k]) + iY_0[j,k]\tan\theta[j,k]}{Y_0[j,k] + i(Y_e[j+1,2k-1] + Y_e[j+1,2k])\tan\theta[j,k]} \right\} \tag{6.8.2}$$

Thus the effective admittance Y_e of the parent tube segment in this bifurcation depends on the characteristic admittance of that segment (Y_0), on the *effective* admittances of the two branch segments ($Y_e[j+1,2k-1]$ and $Y_e[j+1,2k]$), and on the parameter

$$\theta[j,k] = \frac{l[j,k]\,\omega}{c_0[j,k]} \tag{6.8.3}$$

The ultimate aim is to determine the pressure distribution $p_x(x)$ in each tube segment within the tree structure. For this purpose we shall use x repeatedly to measure distance along each tube segment. To avoid confusion we use the notation $x[j,k]$ to identify x in the tube segment located at $[j,k]$, and give it the range $x=0$ to $x=l[j,k]$ where $l[j,k]$ is the length of that segment. The pressure $p_x(x)$ in that tube segment is thus denoted by $p_x(x[j,k])$ and, similarly, as this tube segment divides, the pressure in its branch segments shall be denoted by $p_x(x[j+1,2k-1])$ and $p_x(x[j+1,2k])$.

The pressure distribution in a tube segment at $[j,k]$, using Eq.6.4.11 and present notation, is given by

$$p_x(x[j,k]) = p_0[j,k]$$
$$\times \left\{ e^{-i\omega x[j,k]/c_0[j,k]} + R[j,k]e^{i\omega(x[j,k]-2l[j,k])/c_0[j,k]} \right\} \tag{6.8.4}$$

and similarly, the pressure distribution in a tube segment at $[j-1,n]$ is given by

$$p_x(x[j-1,n]) = p_0[j-1,n]e^{-i\omega x[j-1,n]/c_0[j-1,n]}$$
$$+R[j-1,n]\,p_0[j-1,n]\,e^{i\omega(x[j-1,n]-2l[j-1,n])/c_0[j-1,n]} \tag{6.8.5}$$

If the tube segment at $[j,k]$ is one of the two branch segments of the tube segment at $[j-1,n]$, then

$$n = \frac{k}{2} \quad \text{if } k \text{ is even}$$
$$= \frac{k+1}{2} \quad \text{if } k \text{ is odd} \tag{6.8.6}$$

Furthermore, the pressure in the two tubes must be equal at their junction, that is, at

$$x[j-1,n] = l[j-1,n], \quad x[j,k] = 0 \tag{6.8.7}$$

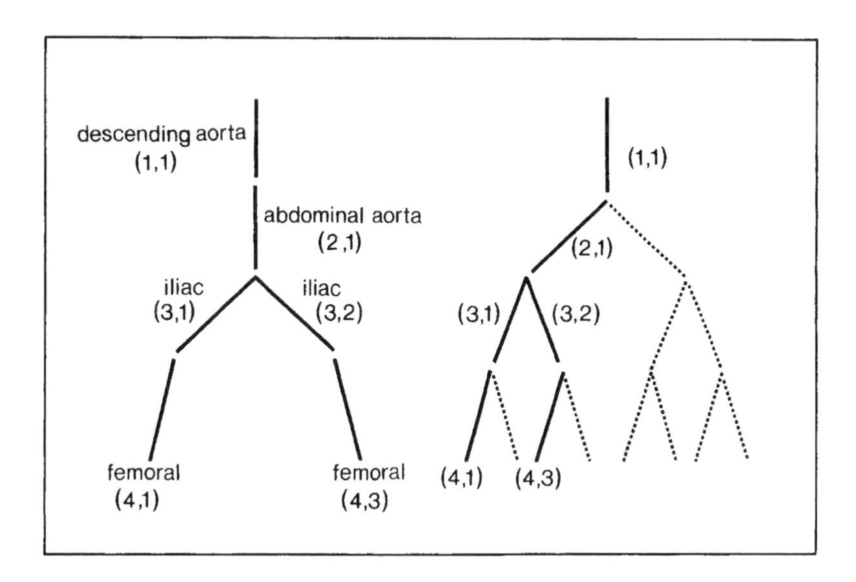

FIGURE 6.8.2. A highly simplified skeleton of the tree structure of the aorta of a dog (left), and the coordinate mapping of its elements (right). The root segment is here denoted by (1,1). In the discussion below it is denoted by [0,0]. From [6].

Evaluating the two pressures at that point we then obtain, from Eqs.6.8.4,5,

$$p_x(0[j,k]) = p_0[j,k]\left\{1 + R[j,k]e^{-2i\omega l[j,k]/c_0[j,k]}\right\} \tag{6.8.8}$$

$$p_x(l[j-1,n]) = p_0[j-1,n] \\ \times\{1 + R[j-1,n]\}e^{-i\omega l[j-1,n]/c_0[j-1,n]} \tag{6.8.9}$$

and equating the two gives

$$p_0[j,k] = p_0[j-1,n]\left\{\frac{(1+R[j-1,n])e^{-i\omega l[j-1,n]/c_0[j-1,n]}}{1+R[j,k]e^{-2i\omega l[j,k]/c_0[j,k]}}\right\} \tag{6.8.10}$$

This result provides an important iterative relation between the amplitudes of the initial forward waves in the two tubes.

Similarly, for the flow wave, writing Eq.6.6.20 in the form

$$q_x(x,t) = q_x(x)e^{i\omega t} \tag{6.8.11}$$

where

$$q_x(x) = q_0\left\{e^{-i\omega x/c_0} - Re^{i\omega(x-2l)/c_0}\right\} \tag{6.8.12}$$

then the flow distribution in a tube at $[j,k]$ is given by

$$q_x(x[j,k]) = q_0[j,k]\,\times$$

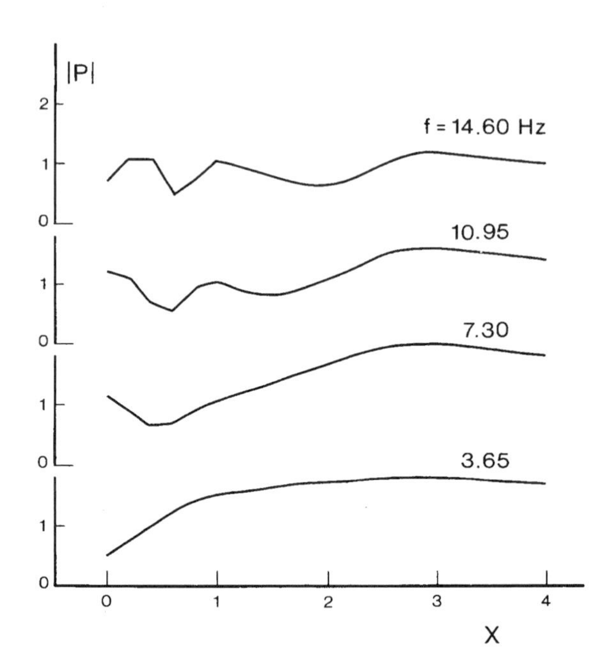

FIGURE 6.8.3. Amplitude of oscillatory pressure wave ($|p|$) as it progresses along the simple tree structure of Fig.6.8.2, at different frequencies. The position coordinate X is normalized at each level of the tree so that the length of each vessel segment is 1.0. From [6].

$$\times \left\{ e^{-i\omega x[j,k]/c_0[j,k]} - R[j,k] e^{i\omega(x[j,k]-2l[j,k])/c_0[j,k]} \right\} \tag{6.8.13}$$

Using Eq.6.6.21 we have

$$q_0[j,k] = Y_0[j,k] p_0[j,k] \tag{6.8.14}$$

and substituting for $p_0[j,k]$ from Eq.6.8.10, we get

$$\begin{aligned} q_0[j,k] &= Y_0[j,k] p_0[j-1,n] \\ &\quad \times \left\{ \frac{(1+R[j-1,n]) e^{-i\omega l[j-1,n]/c_0[j-1,n]}}{1+R[j,k] e^{-2i\omega l[j,k]/c_0[j,k]}} \right\} \end{aligned} \tag{6.8.15}$$

It is seen that p_0 and q_0 provide important links between successive tubes along the tree structure. The values of p_0, q_0 at the downstream end of one tube provide the input values for p_0, q_0 in the next tube, and so on.

The procedure for computing the pressure and flow distributions within the tree structure then begins with a calculation of the effective admittance

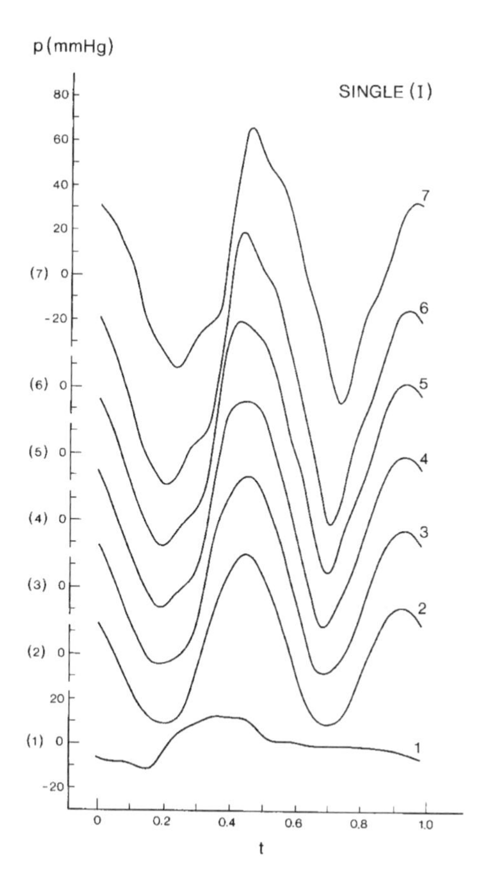

FIGURE 6.8.4. Progression of a composite pressure wave along seven levels of an arterial tree, and the change in its shape because of wave reflections. The pronounced "peaking" seen here has been observed in the human systemic circulation as the cardiac pressure wave progresses down the aorta. From [6].

of each tube segment within the tree structure, using Eq.6.8.2. The equation shows that the effective admittance of a given tube segment depends on the effective admittances of the two branches of that segment, therefore the calculation must proceed from branches to parents, starting at the periphery of the tree structure. At the periphery, the effective admittances of the first set of tube segments must be specified in order to start the calculation. If it can be assumed, for example, that there are no wave reflections from the downstream ends of these tube segments, then their effective admittances are equal to their characteristic admittances, which depend only on tube properties as defined by Eq.6.6.15. Alternatively, if there are wave reflections in these tubes and the reflection coefficients are known or specified, then the effective admittances can be calculated from Eq.6.7.1.

From the calculated effective admittances the reflection coefficients at all junction points can be calculated using Eq.6.8.1. The amplitude p_0 of the initial forward wave in each tube segment is then determined by Eq.6.8.10. In that equation it is seen that the value of p_0 in one tube segment depends on the value of p_0 in its parent tube segment, thus the computation of p_0 must proceed from the root segment of the tree to the periphery. It is convenient to divide all pressures by the value of p_0 in the root segment, namely $p_0[0,0]$ and, similarly, divide all flows by $q_0[0,0]$ and characteristic admittances by $Y_0[0,0]$, using the notation

$$\overline{p}_0[j,k] = \frac{p_0[j,k]}{p_0[0,0]} \tag{6.8.16}$$

$$\begin{aligned}\overline{q}_0[j,k] &= \frac{q_0[j,k]}{q_0[0,0]} = \frac{Y_0[j,k]\,p_0[j,k]}{Y_0[0,0]\,p_0[0,0]} \\ &= \overline{Y}_0[j,k]\overline{p}_0[j,k]\end{aligned} \tag{6.8.17}$$

thus Eqs.6.8.10,15 take the nondimensional form

$$\begin{aligned}\overline{p}_0[j,k] &= \overline{p}_0[j-1,n] \\ &\times \left\{ \frac{(1+R[j-1,n])e^{-i\omega l[j-1,n]/c_0[j-1,n]}}{1+R[j,k]e^{-2i\omega l[j,k]/c_0[j,k]}} \right\}\end{aligned} \tag{6.8.18}$$

$$\begin{aligned}\overline{q}_0[j,k] &= \overline{Y}_0[j,n]\overline{p}_0[j-1,n] \\ &\times \left\{ \frac{(1+R[j-1,n])e^{-i\omega l[j-1,n]/c_0[j-1,n]}}{1+R[j,k]e^{-2i\omega l[j,k]/c_0[j,k]}} \right\}\end{aligned} \tag{6.8.19}$$

These equations can now be used to find $\overline{p}_0, \overline{q}_0$ at successive levels of the tree, starting with the convenient values, from Eqs.6.8.16,17,

$$\overline{p}_0[0,0] = 1.0, \quad \overline{q}_0[0,0] = 1.0 \tag{6.8.20}$$

It is seen from Eqs.6.8.18–20 that in the absence of wave reflections ($R = 0$) these normalized values of the input forward pressure and flow amplitudes will have the value 1.0 throughout the tree structure. This important reference state is useful when wave reflections are present. In that case values of $\overline{p}_0, \overline{q}_0$ in different parts of the tree will deviate from the reference value of 1.0, and these deviations can then be attributed immediately to the effects of wave reflections.

With the computed values of $\overline{p}_0, \overline{q}_0$ throughout the tree structure, the complete expressions for the pressure and flow distributions in each tube segment, from Eqs.6.6.19,20, are now given by

$$\begin{aligned}&\overline{p}(x[j,k],t) = \overline{p}_0[j,k]\; e^{i\omega(t-x[j,k]/c_0[j,k])} \\ &+R[j,k]\;\overline{p}_0[j,k]\; e^{i\omega(t+x[j,k]/c_0[j,k]-2l[j,k]/c_0[j,k])}\end{aligned} \tag{6.8.21}$$

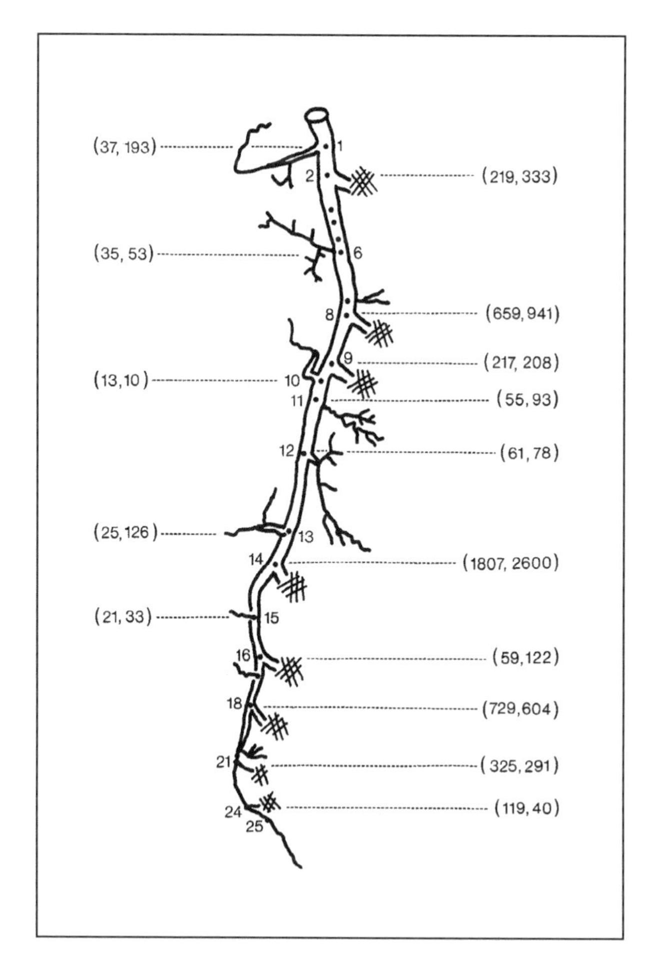

FIGURE 6.8.5. The extensive structure of the right coronary artery of a human heart, obtained from a resin cast of the vessel and its branches. The numbers in brackets indicate the size of the sub-trees arising at different junctions along the main artery. The first number represents the number of branch segment within each sub-tree and the second indicates their collective volume in mm^3. From [21].

$$\overline{q}(x[j,k],t) = \overline{q}_0[j,k]\ e^{i\omega(t - x[j,k]/c_0[j,k])}$$
$$-R[j,k]\ \overline{q}_0[j,k]\ e^{i\omega(t + x[j,k]/c_0[j,k] - 2l[j,k]/c_0[j,k])} \tag{6.8.22}$$

where

$$\overline{p}(x[j,k],t) = \frac{p(x[j,k],t)}{p_0[0,0]} \tag{6.8.23}$$

$$\overline{q}(x[j,k],t) = \frac{q(x[j,k],t)}{q_0[0,0]} \tag{6.8.24}$$

The significance of the pressure and flow distributions along each tube segment and along the tree structure as a whole is seen more clearly by writing Eqs.6.8.21,22 as

$$\overline{p}(x[j,k],t) = \overline{p}_x(x[j,k])e^{i\omega t} \tag{6.8.25}$$

$$\overline{q}(x[j,k],t) = \overline{q}_x(x[j,k])e^{i\omega t} \tag{6.8.26}$$

where $\overline{p}_x(x[j,k])$ and $\overline{q}_x(x[j,k])$ are given by Eqs.6.8.4,13. Since these are generally complex, we introduce the notation

$$\overline{p}(x,t) = \overline{p}_R(x,t) + i\overline{p}_I(x,t) \tag{6.8.27}$$

$$\overline{p}_x(x) = \overline{p}_{xR}(x) + i\overline{p}_{xI}(x) \tag{6.8.28}$$

$$\overline{q}(x,t) = \overline{q}_R(x,t) + i\overline{q}_I(x,t) \tag{6.8.29}$$

$$\overline{q}_x(x) = \overline{q}_{xR}(x) + i\overline{q}_{xI}(x) \tag{6.8.30}$$

then

$$\overline{p}_R(x,t) = \overline{p}_{xR}(x)\cos\omega t - \overline{p}_{xI}(x)\sin\omega t \tag{6.8.31}$$

$$\overline{p}_I(x,t) = \overline{p}_{xI}(x)\cos\omega t + \overline{p}_{xR}(x)\sin\omega t \tag{6.8.32}$$

$$\overline{q}_R(x,t) = \overline{q}_{xR}(x)\cos\omega t - \overline{q}_{xI}(x)\sin\omega t \tag{6.8.33}$$

$$\overline{q}_I(x,t) = \overline{q}_{xI}(x)\cos\omega t + \overline{q}_{xR}(x)\sin\omega t \tag{6.8.34}$$

By simple trigonometry these can be put in the form

$$\overline{p}_R(x,t) = |\overline{p}_x(x)|\cos(\omega t + \delta) \tag{6.8.35}$$

$$\overline{p}_I(x,t) = |\overline{p}_x(x)|\sin(\omega t + \delta) \tag{6.8.36}$$

$$\overline{q}_R(x,t) = |\overline{q}_x(x)|\cos(\omega t + \gamma) \tag{6.8.37}$$

$$\overline{q}_I(x,t) = |\overline{q}_x(x)|\sin(\omega t + \gamma) \tag{6.8.38}$$

where

$$|\overline{p}_x(x)| = \sqrt{\overline{p}_{xR}^2 + \overline{p}_{xI}^2} \tag{6.8.39}$$

$$\tan\delta = \frac{\overline{p}_{xI}}{\overline{p}_{xR}} \tag{6.8.40}$$

$$|\overline{q}_x(x)| = \sqrt{\overline{q}_{xR}^2 + \overline{q}_{xI}^2} \tag{6.8.41}$$

$$\tan\gamma = \frac{\overline{q}_{xI}}{\overline{q}_{xR}} \tag{6.8.42}$$

The absolute value of pressure $|\overline{p}_x(x)|$ thus represents the range of time oscillations at a given position along the tube. *It is the amplitude of time*

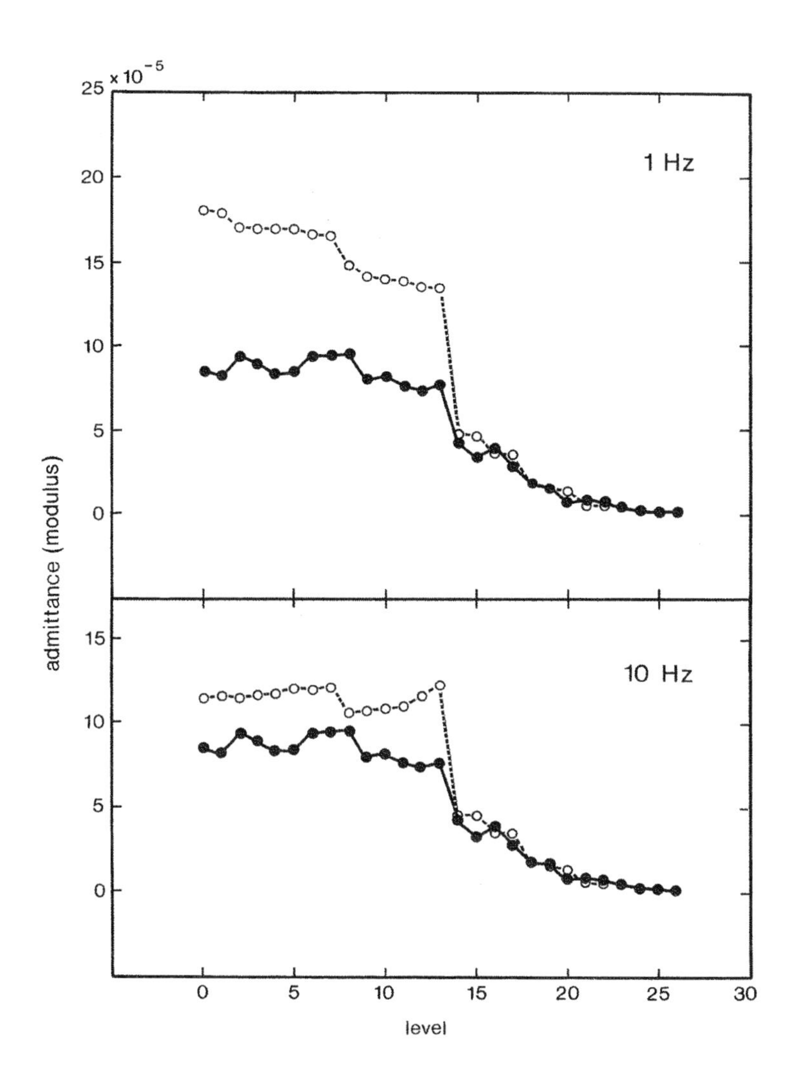

FIGURE 6.8.6. Admittance (in $[cm^3/s]/[dynes/cm^2]$) at different junctions along the main coronary artery depicted in Fig.6.8.5 and at two different frequencies (upper and lower panels). Values of "level" represent the position of each junction along the main artery as numbered in Fig.6.8.5. Solid curve (full circles) represent *characteristic admittance* whose value depends on only static properties of vessel segments, not on frequency. Dashed curve (empty circles) represent *effective admittance*, which depends on wave reflections and hence on frequency. The difference between the two curves in each panel is due entirely to wave reflections. Furthermore, both at high and at low frequencies, effective admittance is seen to be *higher* than characteristic admittance, thus wave reflections are here "helping" the flow by increasing admittance of the propagating wave. This is in contrast with the peaking phenomenon of Fig.6.8.4 where the reverse is true. From [21].

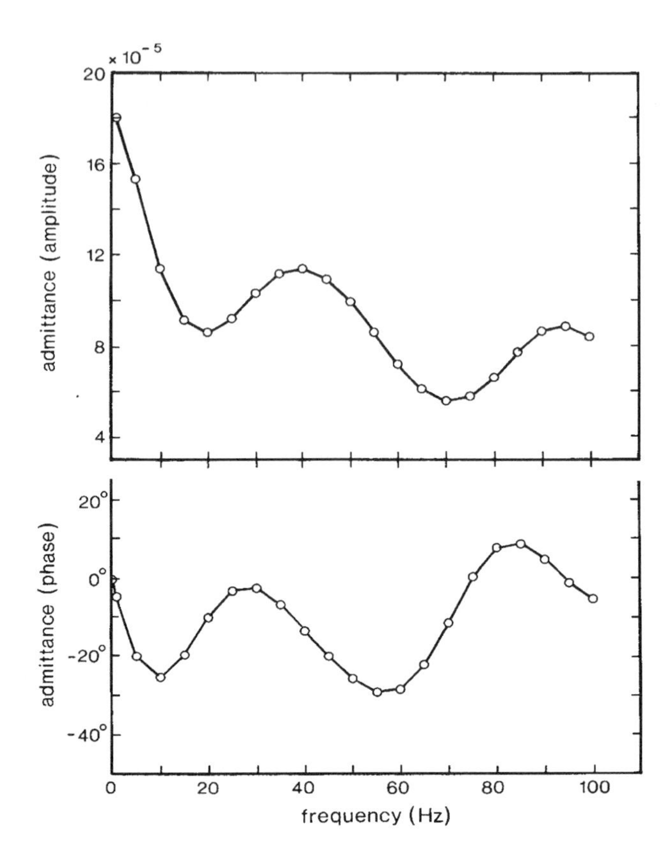

FIGURE 6.8.7. Frequency spectrum of the vascular tree of Fig.6.8.5. Effective admittance is shown here at different frequencies, in amplitude (top) and phase (bottom). The spectrum provides a dynamic profile of the tree, showing how its admittance of pressure and flow waves varies with frequency. The points of maxima and minima, here referred to as "nodes," have a similar interpretation to that of the nodes seen in Fig.6.4.1 for flow in a single tube (see text). From [21].

oscillations at a given position along the tube, thus the distributions of $|\overline{p}_x(x)|$ and of $|\overline{q}_x(x)|$ along the tube are important indicators of the effects of wave reflections. In the absence of wave reflections we have seen that this amplitude has the normalized value of 1.0 everywhere along each tube segment and hence everywhere along the tree. In the presence of wave reflections any deviations from $|\overline{p}_x(x)| = 1.0$ can thus be attributed directly and entirely to wave reflections.

As an example, the pressure distribution along the simple tree structure in Fig.6.8.2, is shown in Fig.6.8.3. A general trend of increasing pressure is seen, which is the cause of the so-called "peaking phenomenon" in the human aorta [6,10,17,19]. The pressure increase observed in Fig.6.8.3 represents an increase in the amplitude of each harmonic of the composite pressure wave generated by the heart as it travels down the aorta. As a result the composite wave becomes "peakier" as illustrated in Fig.6.8.4.

The increase in pressure caused by wave reflections can be viewed as an increase in the impedance of an arterial tree or, equivalently, a decrease in the corresponding admittance of the tree. Recent work based on data from the vascular tree of the human heart [20,21] has shown that wave reflections can also cause a trend of decreasing pressure and hence an *increase* in the admittance of an arterial tree (Figs.6.8.5,6).

The effective admittance and impedance of a vascular tree structure are measures of its dynamic performance. They represent the extent to which a pressure or flow wave entering the tree would be admitted or impeded. They are *dynamic* properties of the tree, which differ in a fundamental way from the characteristic admittance and impedance. They depend on frequency while the characteristic admittance and impedance depend on static properties of the tree only.

For this reason it is necessary to obtain values of effective admittance or impedance of a vascular tree at different frequencies, to produce a so called "frequency spectrum," which is a dyanmic profile of the tree. An example of the admittance spectrum of the vascular tree of Fig.6.8.5 is shown in Fig.6.8.7. The points of maxima and minima on a frequency spectrum are here referred to as "nodes" and their interpretation is similar to, although not as simple as, that of the nodes seen in a single tube (Fig.6.4.1). The first node, for example, is assumed to occur at one quarter wavelength from a major reflection site in analogy with the distance of the first node in a single tube from the reflecting end of the tube (Figs.6.4.1, 6.8.7).

6.9 Problems

1. Is it possible for wave reflections to occur in pulsatile flow in a *rigid* tube? Give reasons for your answer.
2. Explain from the standpoint of physics why wave reflections are important in pulsatile flow.
3. Wave propagation in an elastic tube is made possible by the elasticity of the tube wall, which allows the tube to "bulge out" locally as the pressure increases and to recoil back as the pressure decreases. How then is it possible to consider wave propagation and wave reflections by means of *one-dimensional* analysis in which the radial direction has been eliminated?

4. Show by substitution that the form of $p_x(x)$ in Eq.6.3.6 is the general solution of Eq.6.3.5.
5. Show by substitution that separation of variables in Eq.6.3.4 reduces the *partial* differential equation (Eq.6.2.18) to an *ordinary* differential equation (Eq.6.3.5).
6. Use Eqs.6.4.10,17, with $R = 0.8$, to identify the curve in Fig.6.4.1 that represents the pressure distribution in the tube at one quarter time through the oscillatory cycle, that is, at $\omega t = \pi/2$.
7. Use Eq.6.4.18 to show that the contour of pressure oscillations within the tube, that is, the distribution of $|\overline{p}_x(\overline{x})|$, is as shown in Fig.6.3.1 when there are no wave reflections, and as shown in Fig.6.4.1 when $R = 0.8$.
8. In Eq.6.5.19, if the tube length and wavelength are equal, show that the parameter that embodies the effects of secondary wave reflections, namely, ϵ, is real and find its magnitude if $R_0 = R_l = 0.5$.
9. Find the characteristic admittance of a tube of $1\,cm$ radius, using Eq.6.6.15, if the wave speed is $1\,m/s$ and fluid density is $1\,g/cm^3$. Compare the result with one for a tube of $1\,mm$ radius. Show that the results have the physical dimensions of "flow rate per pressure" as in Eq.6.6.17.
10. Using Eq.6.7.1, show that when the tube length and wavelength are equal, the ratio of effective admittance over characteristic admittance of the tube depends on the reflection coefficient R only. Evaluate this ratio when $R = 0.5$.
11. Consider an arterial bifurcation in isolation, that is, assuming that there are no wave reflections from the downstream end of the two branches, therefore their effective admittances are the same as their characteristic admittances. Show, using Eq.6.7.14, that the reflection coefficient at the junction is then related to the area ratio of the bifurcation. Find its value assuming that the bifurcation is symmetrical and that the cube law is in effect.
12. Consider the *characteristic* admittance at different levels in a tree structure. If the branch diameters in the tree are assumed to obey the cube law, does the characteristic admittance increase or decrease from one level of the tree to the next? In the presence of wave reflections, of course, the actual admittance at each level is the *effective* admittance. How would you expect the *effective* admittance to change from one level of the tree to the next?

6.10 References and Further Reading

1. Morgan GW, Kiely JP, 1954. Wave propagation in a viscous liquid contained in a flexible tube. Journal of Acoustical Society of America 26:323–328.

2. Womersley JR, 1957. Oscillatory flow in arteries: the constrained elastic tube as a model of arterial flow and pulse transmission. Physics in Medicine and Biology 2:178–187.

3. Patel DJ, Fry DL, 1966. Longitudinal tethering of arteries in dogs. Circulation Research 19:1011–1021.

4. Duan B, Zamir M, 1993. Reflection coefficients in pulsatile flow through converging junctions and the pressure distribution in a simple loop. Journal of Biomechanics 26:1439–1447.

5. Duan B, Zamir M, 1995. Mechanics of wave reflections in a coronary bypass loop model: The possibility of partial flow cut-off. Journal of Biomechanics 28:567–574.

6. Duan B, Zamir M, 1995. Pressure peaking in pulsatile flow through arterial tree structures. Annals of Biomedical Engineering 23:794–803.

7. Taylor MG, 1957. An approach to the analysis of the arterial pulse wave. I. Oscillations in an attenuating line. Physics in Medicine and Biology I:258–269.

8. Taylor MG, 1966. The input impedance of an assembly of randomly branching tubes. Biophysical Journal 6:29–51.

9. Wylie EB, Streeter V, 1978. Fluid Transient. McGraw-Hill, New York.

10. Fung YC, 1984. Biodynamics: Circulation. Springer-Verlag, New York.

11. Hardung V, 1962. Propagation of pulse waves in viscoelastic tubings. Handbook of Physiology: Circulation. 1:107–135. Williams and Wilkins, Baltimore.

12. Lighthill MJ, 1975. Mathematical Biofluiddynamics. Society for Industrial and Applied Mathematics, Philadelphia.

13. Duan B, Zamir M, 1992. Viscous damping in one-dimensional wave transmission. Journal of Acoustical Society of America 92:3358–3363.

14. Spiegel MR, 1967. Applied Differential Equations. Prentice Hall, Englewood Cliffs, New Jersey.

15. Kenner T, 1969. The dynamics of pulsatile flow in the coronary arteries. Pflügers Archiv; European Journal of Physiology 310:22–34.

16. Spiegel MR, 1968. Mathematical Handbook of Formulas and Tables. McGraw-Hill, New York.

17. McDonald DA, 1974. Blood flow in arteries. Edward Arnold, London.

18. Milnor WR, 1989. Hemodynamics. Williams and Wilkins, Baltimore.

19. Kouchoukos NT, Sheppard LC, Mcdonald DA, 1970. Estimation of stroke volume in the dog by pulse contour method. Circulation Research 26:611–623.

20. Zamir M, 1996. Tree structure and branching characteristics of the right coronary artery in a right-dominant heart. Canadian Journal of Cardiology 12(6):593–599.

21. Zamir M, 1998. Mechanics of blood supply to the heart: wave reflection effects in a right coronary artery. Proceedings of the Royal Society of London B265:439–444.

Appendix A
Values of $\boldsymbol{J}_0(\boldsymbol{\Lambda})$, $\boldsymbol{J}_1(\boldsymbol{\Lambda})$, and $\boldsymbol{G}(\boldsymbol{\Lambda})$

Ω	Λ	$J_0(\Lambda)$
0.0	0.0000+0.0000i	1.0000+0.0000i
0.1	-0.0707+0.0707i	1.0000+0.0025i
0.2	-0.1414+0.1414i	1.0000+0.0100i
0.3	-0.2121+0.2121i	0.9999+0.0225i
0.4	-0.2828+0.2828i	0.9996+0.0400i
0.5	-0.3536+0.3536i	0.9990+0.0625i
0.6	-0.4243+0.4243i	0.9980+0.0900i
0.7	-0.4950+0.4950i	0.9962+0.1224i
0.8	-0.5657+0.5657i	0.9936+0.1599i
0.9	-0.6364+0.6364i	0.9898+0.2023i
1.0	-0.7071+0.7071i	0.9844+0.2496i
1.1	-0.7778+0.7778i	0.9771+0.3017i
1.2	-0.8485+0.8485i	0.9676+0.3587i
1.3	-0.9192+0.9192i	0.9554+0.4204i
1.4	-0.9899+0.9899i	0.9401+0.4867i
1.5	-1.0607+1.0607i	0.9211+0.5576i
1.6	-1.1314+1.1314i	0.8979+0.6327i
1.7	-1.2021+1.2021i	0.8700+0.7120i
1.8	-1.2728+1.2728i	0.8367+0.7953i
1.9	-1.3435+1.3435i	0.7975+0.8821i

Ω	Λ	$J_0(\Lambda)$
2.0	-1.4142+1.4142i	0.7517+0.9723i
2.1	-1.4849+1.4849i	0.6987+1.0654i
2.2	-1.5556+1.5556i	0.6377+1.1610i
2.3	-1.6263+1.6263i	0.5680+1.2585i
2.4	-1.6971+1.6971i	0.4890+1.3575i
2.5	-1.7678+1.7678i	0.4000+1.4572i
2.6	-1.8385+1.8385i	0.3001+1.5569i
2.7	-1.9092+1.9092i	0.1887+1.6557i
2.8	-1.9799+1.9799i	0.0651+1.7529i
2.9	-2.0506+2.0506i	-0.0714+1.8472i
3.0	-2.1213+2.1213i	-0.2214+1.9376i
3.1	-2.1920+2.1920i	-0.3855+2.0228i
3.2	-2.2627+2.2627i	-0.5644+2.1016i
3.3	-2.3335+2.3335i	-0.7584+2.1723i
3.4	-2.4042+2.4042i	-0.9680+2.2334i
3.5	-2.4749+2.4749i	-1.1936+2.2832i
3.6	-2.5456+2.5456i	-1.4353+2.3199i
3.7	-2.6163+2.6163i	-1.6933+2.3413i
3.8	-2.6870+2.6870i	-1.9674+2.3454i
3.9	-2.7577+2.7577i	-2.2576+2.3300i
4.0	-2.8284+2.8284i	-2.5634+2.2927i
4.1	-2.8991+2.8991i	-2.8843+2.2309i
4.2	-2.9698+2.9698i	-3.2195+2.1422i
4.3	-3.0406+3.0406i	-3.5679+2.0236i
4.4	-3.1113+3.1113i	-3.9283+1.8726i
4.5	-3.1820+3.1820i	-4.2991+1.6860i
4.6	-3.2527+3.2527i	-4.6784+1.4610i
4.7	-3.3234+3.3234i	-5.0639+1.1946i
4.8	-3.3941+3.3941i	-5.4531+0.8837i
4.9	-3.4648+3.4648i	-5.8429+0.5251i
5.0	-3.5355+3.5355i	-6.2301+0.1160i
5.1	-3.6062+3.6062i	-6.6107-0.3467i
5.2	-3.6770+3.6770i	-6.9803-0.8658i
5.3	-3.7477+3.7477i	-7.3344-1.4443i
5.4	-3.8184+3.8184i	-7.6674-2.0845i
5.5	-3.8891+3.8891i	-7.9736-2.7890i
5.6	-3.9598+3.9598i	-8.2466-3.5597i
5.7	-4.0305+4.0305i	-8.4794-4.3986i
5.8	-4.1012+4.1012i	-8.6644-5.3068i
5.9	-4.1719+4.1719i	-8.7937-6.2854i

Ω	Λ	$J_0(\Lambda)$
6.0	-4.2426+4.2426i	-8.8583-7.3347i
6.1	-4.3134+4.3134i	-8.8491-8.4545i
6.2	-4.3841+4.3841i	-8.7561-9.6437i
6.3	-4.4548+4.4548i	-8.5688-10.9007i
6.4	-4.5255+4.5255i	-8.2762-12.2229i
6.5	-4.5962+4.5962i	-7.8669-13.6065i
6.6	-4.6669+4.6669i	-7.3287-15.0470i
6.7	-4.7376+4.7376i	-6.6492-16.5384i
6.8	-4.8083+4.8083i	-5.8155-18.0736i
6.9	-4.8790+4.8790i	-4.8146-19.6440i
7.0	-4.9497+4.9497i	-3.6329-21.2394i
7.1	-5.0205+5.0205i	-2.2571-22.8481i
7.2	-5.0912+5.0912i	-0.6737-24.4565i
7.3	-5.1619+5.1619i	1.1308-26.0492i
7.4	-5.2326+5.2326i	3.1695-27.6088i
7.5	-5.3033+5.3033i	5.4550-29.1157i
7.6	-5.3740+5.3740i	7.9994-30.5483i
7.7	-5.4447+5.4447i	10.8140-31.8824i
7.8	-5.5154+5.5154i	13.9089-33.0915i
7.9	-5.5861+5.5861i	17.2931-34.1468i
8.0	-5.6569+5.6569i	20.9740-35.0167i
8.1	-5.7276+5.7276i	24.9569-35.6671i
8.2	-5.7983+5.7983i	29.2452-36.0611i
8.3	-5.8690+5.8690i	33.8398-36.1594i
8.4	-5.9397+5.9397i	38.7384-35.9198i
8.5	-6.0104+6.0104i	43.9359-35.2977i
8.6	-6.0811+6.0811i	49.4231-34.2458i
8.7	-6.1518+6.1518i	55.1869-32.7143i
8.8	-6.2225+6.2225i	61.2097-30.6514i
8.9	-6.2933+6.2933i	67.4687-28.0029i
9.0	-6.3640+6.3640i	73.9357-24.7128i
9.1	-6.4347+6.4347i	80.5764-20.7236i
9.2	-6.5054+6.5054i	87.3500-15.9764i
9.3	-6.5761+6.5761i	94.2084-10.4117i
9.4	-6.6468+6.6468i	101.0964-3.9693i
9.5	-6.7175+6.7175i	107.9500+3.4106i
9.6	-6.7882+6.7882i	114.6971+11.7870i
9.7	-6.8589+6.8589i	121.2561+21.2175i
9.8	-6.9296+6.9296i	127.5357+31.7575i
9.9	-7.0004+7.0004i	133.4345+43.4592i

Ω	Λ	$J_0(\Lambda)$
10.0	-7.0711+7.0711i	138.8405+56.3705i
10.1	-7.1418+7.1418i	143.6306+70.5343i
10.2	-7.2125+7.2125i	147.6705+85.9873i
10.3	-7.2832+7.2832i	150.8141+102.7583i
10.4	-7.3539+7.3539i	152.9034+120.8673i
10.5	-7.4246+7.4246i	153.7686+140.3238i
10.6	-7.4953+7.4953i	153.2277+161.1250i
10.7	-7.5660+7.5660i	151.0869+183.2547i
10.8	-7.6368+7.6368i	147.1407+206.6808i
10.9	-7.7075+7.7075i	141.1724+231.3540i
11.0	-7.7782+7.7782i	132.9544+257.2052i
11.1	-7.8489+7.8489i	122.2493+284.1439i
11.2	-7.9196+7.9196i	108.8104+312.0556i
11.3	-7.9903+7.9903i	92.3834+340.7999i
11.4	-8.0610+8.0610i	72.7073+370.2078i
11.5	-8.1317+8.1317i	49.5166+400.0798i
11.6	-8.2024+8.2024i	22.5427+430.1828i
11.7	-8.2731+8.2731i	-8.4832+460.2484i
11.8	-8.3439+8.3439i	-43.8279+489.9697i
11.9	-8.4146+8.4146i	-83.7530+518.9998i
12.0	-8.4853+8.4853i	-128.5116+546.9486i
12.1	-8.5560+8.5560i	-178.3446+573.3811i
12.2	-8.6267+8.6267i	-233.4761+597.8151i
12.3	-8.6974+8.6974i	-294.1095+619.7195i
12.4	-8.7681+8.7681i	-360.4215+638.5122i
12.5	-8.8388+8.8388i	-432.5575+653.5589i
12.6	-8.9095+8.9095i	-510.6247+664.1719i
12.7	-8.9803+8.9803i	-594.6859+669.6094i
12.8	-9.0510+9.0510i	-684.7528+669.0752i
12.9	-9.1217+9.1217i	-780.7779+661.7186i
13.0	-9.1924+9.1924i	-882.6466+646.6356i
13.1	-9.2631+9.2631i	-990.1690+622.8702i
13.2	-9.3338+9.3338i	-1103.0706+589.4163i
13.3	-9.4045+9.4045i	-1220.9828+545.2208i
13.4	-9.4752+9.4752i	-1343.4335+489.1877i
13.5	-9.5459+9.5459i	-1469.8363+420.1827i
13.6	-9.6167+9.6167i	-1599.4804+337.0393i
13.7	-9.6874+9.6874i	-1731.5195+238.5659i
13.8	-9.7581+9.7581i	-1864.9609+123.5544i
13.9	-9.8288+9.8288i	-1998.6544-9.2100i

Ω	Λ	$J_0(\Lambda)$
14.0	-9.8995+9.8995i	-2131.2812-160.9377i
14.1	-9.9702+9.9702i	-2261.3426-332.8211i
14.2	-10.0409+10.0409i	-2387.1497-526.0207i
14.3	-10.1116+10.1116i	-2506.8122-741.6479i
14.4	-10.1823+10.1823i	-2618.2294-980.7472i
14.5	-10.2530+10.2530i	-2719.0803-1244.2754i
14.6	-10.3238+10.3238i	-2806.8153-1533.0796i
14.7	-10.3945+10.3945i	-2878.6497-1847.8721i
14.8	-10.4652+10.4652i	-2931.5570-2189.2041i
14.9	-10.5359+10.5359i	-2962.2652-2557.4363i
15.0	-10.6066+10.6066i	-2967.2545-2952.7079i

Ω	$J_1(\Lambda)$	$G(\Lambda)$
0.0	0.0000+0.0000i	1.0000-0.0000i
0.1	-0.0354+0.0353i	1.0000-0.0000i
0.2	-0.0711+0.0704i	1.0000-0.0000i
0.3	-0.1073+0.1049i	1.0000-0.0000i
0.4	-0.1442+0.1386i	1.0001-0.0000i
0.5	-0.1822+0.1712i	1.0003-0.0000i
0.6	-0.2215+0.2024i	1.0007-0.0001i
0.7	-0.2623+0.2320i	1.0012-0.0002i
0.8	-0.3049+0.2596i	1.0021-0.0004i
0.9	-0.3493+0.2849i	1.0033-0.0008i
1.0	-0.3959+0.3076i	1.0049-0.0014i
1.1	-0.4447+0.3272i	1.0071-0.0025i
1.2	-0.4959+0.3435i	1.0097-0.0042i
1.3	-0.5496+0.3559i	1.0127-0.0066i
1.4	-0.6059+0.3642i	1.0162-0.0101i
1.5	-0.6649+0.3679i	1.0198-0.0147i
1.6	-0.7264+0.3664i	1.0234-0.0208i
1.7	-0.7905+0.3594i	1.0265-0.0285i
1.8	-0.8571+0.3463i	1.0287-0.0376i
1.9	-0.9260+0.3266i	1.0295-0.0482i
2.0	-0.9971+0.2998i	1.0285-0.0598i
2.1	-1.0700+0.2653i	1.0252-0.0718i
2.2	-1.1445+0.2225i	1.0197-0.0837i
2.3	-1.2202+0.1708i	1.0121-0.0947i
2.4	-1.2966+0.1098i	1.0027-0.1043i
2.5	-1.3731+0.0387i	0.9922-0.1123i
2.6	-1.4491-0.0430i	0.9809-0.1184i
2.7	-1.5240-0.1359i	0.9695-0.1227i
2.8	-1.5968-0.2406i	0.9583-0.1254i
2.9	-1.6667-0.3576i	0.9476-0.1268i
3.0	-1.7326-0.4875i	0.9375-0.1271i
3.1	-1.7935-0.6306i	0.9281-0.1266i
3.2	-1.8481-0.7875i	0.9194-0.1254i
3.3	-1.8949-0.9585i	0.9115-0.1238i
3.4	-1.9327-1.1440i	0.9043-0.1219i
3.5	-1.9596-1.3440i	0.8977-0.1198i
3.6	-1.9742-1.5589i	0.8917-0.1175i
3.7	-1.9744-1.7885i	0.8862-0.1152i
3.8	-1.9584-2.0327i	0.8811-0.1128i
3.9	-1.9241-2.2913i	0.8765-0.1105i

Ω	$J_1(\Lambda)$	$G(\Lambda)$
4.0	-1.8692-2.5638i	0.8722-0.1082i
4.1	-1.7916-2.8496i	0.8682-0.1060i
4.2	-1.6886-3.1479i	0.8645-0.1039i
4.3	-1.5579-3.4576i	0.8611-0.1018i
4.4	-1.3969-3.7774i	0.8579-0.0998i
4.5	-1.2028-4.1057i	0.8549-0.0978i
4.6	-0.9730-4.4406i	0.8520-0.0960i
4.7	-0.7046-4.7801i	0.8493-0.0942i
4.8	-0.3949-5.1214i	0.8468-0.0925i
4.9	-0.0410-5.4618i	0.8444-0.0908i
5.0	0.3598-5.7979i	0.8421-0.0892i
5.1	0.8102-6.1260i	0.8399-0.0877i
5.2	1.3128-6.4421i	0.8378-0.0862i
5.3	1.8701-6.7414i	0.8358-0.0847i
5.4	2.4844-7.0189i	0.8338-0.0834i
5.5	3.1579-7.2690i	0.8320-0.0820i
5.6	3.8922-7.4857i	0.8302-0.0807i
5.7	4.6889-7.6622i	0.8285-0.0795i
5.8	5.5492-7.7914i	0.8269-0.0782i
5.9	6.4736-7.8657i	0.8253-0.0770i
6.0	7.4622-7.8767i	0.8238-0.0759i
6.1	8.5146-7.8156i	0.8223-0.0748i
6.2	9.6296-7.6730i	0.8209-0.0737i
6.3	10.8053-7.4391i	0.8195-0.0726i
6.4	12.0389-7.1035i	0.8182-0.0716i
6.5	13.3267-6.6553i	0.8169-0.0706i
6.6	14.6639-6.0832i	0.8157-0.0696i
6.7	16.0446-5.3755i	0.8145-0.0686i
6.8	17.4616-4.5201i	0.8133-0.0677i
6.9	18.9063-3.5048i	0.8122-0.0668i
7.0	20.3689-2.3172i	0.8111-0.0659i
7.1	21.8377-0.9445i	0.8100-0.0651i
7.2	23.2995+0.6256i	0.8090-0.0642i
7.3	24.7391+2.4056i	0.8080-0.0634i
7.4	26.1397+4.4075i	0.8071-0.0626i
7.5	27.4822+6.6429i	0.8061-0.0618i
7.6	28.7456+9.1229i	0.8052-0.0610i
7.7	29.9065+11.8575i	0.8043-0.0603i
7.8	30.9394+14.8559i	0.8035-0.0596i
7.9	31.8163+18.1258i	0.8026-0.0589i

Ω	$J_1(\Lambda)$	$G(\Lambda)$
8.0	32.5069+21.6735i	0.8018-0.0582i
8.1	32.9782+25.5034i	0.8010-0.0575i
8.2	33.1950+29.6177i	0.8002-0.0568i
8.3	33.1195+34.0162i	0.7995-0.0562i
8.4	32.7112+38.6959i	0.7988-0.0555i
8.5	31.9274+43.6505i	0.7980-0.0549i
8.6	30.7229+48.8702i	0.7973-0.0543i
8.7	29.0503+54.3412i	0.7967-0.0537i
8.8	26.8601+60.0452i	0.7960-0.0531i
8.9	24.1008+65.9591i	0.7954-0.0526i
9.0	20.7192+72.0543i	0.7947-0.0520i
9.1	16.6608+78.2964i	0.7941-0.0515i
9.2	11.8699+84.6448i	0.7935-0.0509i
9.3	6.2902+91.0518i	0.7929-0.0504i
9.4	-0.1349+97.4626i	0.7923-0.0499i
9.5	-7.4614+103.8144i	0.7918-0.0494i
9.6	-15.7448+110.0360i	0.7912-0.0489i
9.7	-25.0385+116.0475i	0.7907-0.0484i
9.8	-35.3939+121.7595i	0.7901-0.0479i
9.9	-46.8590+127.0730i	0.7896-0.0475i
10.0	-59.4776+131.8786i	0.7891-0.0470i
10.1	-73.2883+136.0567i	0.7886-0.0465i
10.2	-88.3234+139.4764i	0.7881-0.0461i
10.3	-104.6075+141.9960i	0.7877-0.0457i
10.4	-122.1564+143.4624i	0.7872-0.0453i
10.5	-140.9753+143.7110i	0.7867-0.0448i
10.6	-161.0577+142.5659i	0.7863-0.0444i
10.7	-182.3835+139.8400i	0.7859-0.0440i
10.8	-204.9172+135.3350i	0.7854-0.0436i
10.9	-228.6061+128.8422i	0.7850-0.0432i
11.0	-253.3784+120.1428i	0.7846-0.0429i
11.1	-279.1414+109.0086i	0.7842-0.0425i
11.2	-305.7787+95.2032i	0.7838-0.0421i
11.3	-333.1488+78.4831i	0.7834-0.0418i
11.4	-361.0824+58.5988i	0.7830-0.0414i
11.5	-389.3800+35.2971i	0.7826-0.0410i
11.6	-417.8102+8.3226i	0.7823-0.0407i
11.7	-446.1067-22.5802i	0.7819-0.0404i
11.8	-473.9663-57.6635i	0.7815-0.0400i
11.9	-501.0465-97.1742i	0.7812-0.0397i

Ω	$J_1(\Lambda)$	$G(\Lambda)$
12.0	-526.9634-141.3498i	0.7808-0.0394i
12.1	-551.2897-190.4153i	0.7805-0.0391i
12.2	-573.5524-244.5786i	0.7802-0.0388i
12.3	-593.2310-304.0262i	0.7798-0.0384i
12.4	-609.7565-368.9184i	0.7795-0.0381i
12.5	-622.5092-439.3834i	0.7792-0.0378i
12.6	-630.8185-515.5116i	0.7789-0.0376i
12.7	-633.9615-597.3494i	0.7786-0.0373i
12.8	-631.1637-684.8921i	0.7783-0.0370i
12.9	-621.5984-778.0766i	0.7780-0.0367i
13.0	-604.3881-876.7738i	0.7777-0.0364i
13.1	-578.6061-980.7798i	0.7774-0.0362i
13.2	-543.2783-1089.8078i	0.7771-0.0359i
13.3	-497.3866-1203.4786i	0.7768-0.0356i
13.4	-439.8726-1321.3110i	0.7765-0.0354i
13.5	-369.6429-1442.7120i	0.7763-0.0351i
13.6	-285.5745-1566.9669i	0.7760-0.0349i
13.7	-186.5226-1693.2281i	0.7757-0.0346i
13.8	-71.3285-1820.5053i	0.7755-0.0344i
13.9	61.1703-1947.6544i	0.7752-0.0341i
14.0	212.1289-2073.3667i	0.7750-0.0339i
14.1	382.6821-2196.1588i	0.7747-0.0336i
14.2	573.9303-2314.3615i	0.7745-0.0334i
14.3	786.9226-2426.1106i	0.7742-0.0332i
14.4	1022.6393-2529.3371i	0.7740-0.0330i
14.5	1281.9715-2621.7587i	0.7738-0.0327i
14.6	1565.6992-2700.8717i	0.7735-0.0325i
14.7	1874.4672-2763.9449i	0.7733-0.0323i
14.8	2208.7588-2808.0135i	0.7731-0.0321i
14.9	2568.8672-2829.8764i	0.7728-0.0319i
15.0	2954.8653-2826.0936i	0.7726-0.0317i

Values of $G(\Lambda)$ are based on a ratio of wall thickness to tube radius of 0.1, a ratio of wall density to fluid density of 1.0, and a Poisson's ratio of 0.5.

Appendix B
Solutions to Problems

Chapter 1

1. Sand is not a fluid since it is not a continuum on the macroscopic scale. That is, sand is granular on the macroscopic scale and its grains are not continuous with each other. When sand is poured, its grains move under the action of gravity, *independently from each other*, except for occasional collisions or rolling over each other. Blood, by comparison, is composed of corpuscular bodies within a continuous plasma. The size of the corpuscles is of the order of $10^{-5}\,m$, while grains of sand are visible to the naked eye and are 10 to 100 times larger. More important, the plasma within which the corpuscles of blood are immersed is continuous on the macroscopic scale and it gives blood its fluid character. The presence of corpuscles modifies only the details of that character.

2. If the size of a blood corpuscle is taken as 10 microns, then 2500 corpuscles can fit within the diameter of the aorta, which means that the scale of blood flow within the aorta is much larger than the scale of blood corpuscles. Under these circumstances, it is reasonable to consider blood as a homogeneous fluid and to consider blood flow in the aorta as flow of a homogeneous Newtonian fluid. In vessels of smaller diameter this model will become gradually less valid until the two scales become comparable, which occurs in the capillaries. Blood within the capillaries cannot be considered a homogeneous fluid, indeed flow within the capillaries cannot be regarded as fluid flow in the same way as that in the aorta.

3. A fluid body moves in bulk, as a solid body, when the forces causing it to move act uniformly, that is, act equally on every element of the body, and when every element of the body is equally free or constrained in its motion. This high degree of uniformity is rarely met in practice. Fluid in a cup can be in motion, with the cup, but not flowing, if the cup is moved carefully with constant velocity so as not to "disturb" the fluid within. Motion of the cup exerts a uniform force on the body of the fluid, causing all its elements to move in unison with each other and with the cup. Disturbing the fluid within the cup means disturbing the uniformity of this force, which can come about if the cup is shaken or accelerated. The situation has an important parallel in the vascular system. Blood contained within the vessels is generally unaffected by reasonably slow motion of the body housing these vessels. But if the body is violently shaken or subjected to high acceleration or impact, the action may trigger violent flow within the vessels to the point of damaging or rupturing the vessels.

4. The laws of fluid motion require the velocities of specific elements of a fluid body, that is, they require a so-called "Lagrangian" description of the flow field. In this description, however, each element of fluid must be labeled and followed, individually, as in the motion of a single particle. The impracticality of doing so is the reason for using Eulerian velocities instead. The way in which the laws of motion are adapted to Eulerian velocities instead of Lagrangian velocities is outlined in the derivation of the equations of fluid flow, which is the main subject of Chapter 2.

5. The acceleration in a flow field, in terms of *Eulerian* velocities, depends on time derivatives as well as space derivatives of the velocities (Eq.1.6.5). Therefore:

a. False, since the velocity gradients in space may not be zero.
b. False, since the velocity gradients in time may not be zero.
c. False, since a steady flow field is one in which the (Eulerian) velocities are not functions of time, hence the same as 'a' above.
d. False, since a uniform flow field is one in which the (Eulerian) velocities are not functions of position, hence the same as 'b' above.

6. A blood cell immersed within the plasma of blood is not a legitimate "element" of the continuum representing the blood as whole, as a fluid body. Thus a single cell may engage in motion that is not part of the fluid motion of the continuum within the vessel. In particular, near the vessel wall, the cell may skid on a thin layer of plasma between it and the vessel wall, the plasma acting as a lubricant. The no-slip boundary condition at the vessel wall in this case applies to the elements of *plasma* in contact with the vessel wall, not the skidding cell, which is not in contact with the wall. If the cell does come in contact with the vessel wall, the no-slip boundary condition will arrest the part of the cell in contact with the wall while the rest of the cell continues to be swept by the moving plasma away from the wall. The result of this scenario would be "rolling" of the blood cell, a phenomenon that has also been observed to occur near the vessel wall.

7. The Reynolds number R for flow in a tube is defined by (Eq.1.9.1)

$$R = \frac{\rho \bar{u} d}{\mu}$$

where $\overline{u}$ is the average flow velocity through the tube, ρ,μ are the density and viscosity of the fluid, and d is diameter of the tube. According to Reynolds original experiments, at $R \approx 2000$ laminar flow in a tube became unstable and a transition occurred from laminar to turbulent flow. Later it became apparent that $R = 2000$ is only a lower bound for transition to turbulent flow. That is, at values of R less than 2000 flow in a tube would normally be laminar, and any disturbances within the flow would decay rather than lead to turbulent flow. At values of R above 2000, disturbances may amplify and lead to turbulent flow, the exact value in each case depends on a number of other factors such as the nature of the disturbances and roughness of the tube wall.
In the cardiovascular system, at a normal systemic blood flow rate the value of R in the ascending aorta would be near 1000 (Eq.1.9.3). In smaller branches of the aorta values of R would be lower because of smaller diameters as well as lower velocities. Under these *normal* circumstances, therefore, flow would be everywhere laminar. At higher levels of activity, however, systemic blood flow increases considerably and values of R well in excess of 2000 may be reached. Also, turbulent flow can and does occur outside these normal circumstances, usually because of a local increase in flow rate when a blood vessel is narrowed because of a lesion, or when flow through a valve is disrupted because of a lesion in the valve leaflets. In both cases the local presence (and sound) of turbulent flow is an important clinical clue to a potential pathology. In *pulsatile* flow the onset of turbulence is further complicated by an added time dimension to the problem, with flow rate and velocity reaching their peaks only momentarily within the oscillatory cycle.

Chapter 2

1. The equations of fluid flow are based on the law of conservation of mass, and on Newton's law of motion, which requires that the product of mass and acceleration of an object in motion be equal to the net force acting on it. The first yields the so-called equation of continuity (Eq.2.4.8), and the second yields the Navier–Stokes equations (Eqs.2.8.1–3).

2. When a fluid body is in motion, its elements do not move in unison, hence the equations of fluid flow must deal with each element *individu-*

ally. Because on the macroscopic scale each fluid element is represented by a "point," a *material point*, the equations must be applied at every point of the fluid body. They are thus "point equations" in the sense that they do not apply to the body as a whole but at every point within it.

3. When the density of a given mass of fluid is constant, the conservation of its mass becomes identical with the conservation of its volume. The equation of continuity (Eq.2.4.8) prescribes certain conditions that velocity gradients at a given point in a flow field must satisfy in order to ensure conservation of volume of an element of fluid at that point. Velocity gradients are involved because they represent rates of deformation of the volume of a fluid element.

4. The Navier–Stokes equations (Eqs.2.8.1–3) are based on Newton's law of motion, namely (Eq.2.5.1)

$$m\,a = F$$

where m is the mass of an object in motion, a is its acceleration, and F is the force acting on it, both a and F being vector quantities. When this law is applied to a fluid element, the mass of the element is eliminated by dividing both sides of the equation by the volume v of the element so that the equation becomes

$$\rho\,a = f$$

where $\rho = m/v$ is the density of the element and $f = F/v$ is the force acting on it *per unit volume*. As seen in Chapter 2, both a and f can then be expressed in terms of other flow properties without actually requiring either the mass m or the volume v. The *identity* of the fluid element to which the equations are applied is actually determined by its coordinate position (x, r, θ) at time t. Indeed the equation more fully reads

$$\rho(x, r, \theta, t)\; a(x, r, \theta, t) = f(x, r, \theta, t)$$

Each "point" in space and time, that is, each set of values x, r, θ, t, identifies an individual fluid element, and applies the equation of motion to that particular element.

5. A body force acts directly on the mass of a fluid element in motion and can reach it from a distance without being in contact with it. A boundary force acts on the boundary of the element and can reach it only by contact with neighboring elements. An example of the first is gravitational force and an example of the second is pressure or shear force. In the absense of gravitational forces, flow in a tube is governed by a balance of pressure and shear forces. The first acts as the driving force and the second as a retarding force caused by the viscosity of the fluid. A gravitational force may add to the driving or retarding force, depending on whether flow is in the direction of gravity or against it, as in the case of flow in an inclined tube.

6. The shear stress acting at the tube wall to the effect of opposing the flow in a tube is the component of the stress tensor that acts on coordinate surfaces perpendicular to r and is directed in the x direction, namely, τ_{rx}. It is related to the velocity gradients in that plane by (Eq.2.7.2)

$$\tau_{rx} = \mu \left(\frac{\partial u}{\partial r} + \frac{\partial v}{\partial x} \right)$$

The relation between the stress and velocity gradients is linear, hence it is based on the assumption that the fluid is Newtonian. Furthermore, if the radial velocity v is zero, as it is in steady or pulsatile flow in a rigid tube, or sufficiently small to be negligible as in the case of pulsatile flow in an elastic tube when the propagating wave length is much larger than the tube radius, this component of the shear stress takes on the simpler form (Eqs.5.5.2,3)

$$\tau_{rx} = \mu \frac{\partial u}{\partial r}$$

7. Eq.2.8.1 as it stands has the form "*density* $\times$ *acceleration* + *pressure force* = *viscous force*" in which the pressure and viscous forces represent *forces per unit volume.* Thus multiplying the equation through by the volume of a fluid element and combining the two forces gives the required form.

8. The Navier–Stokes equations and equation of continuity in rectangular Cartesian coordinates x, y, z and corresponding velocity components u, v, w are given by

$$\rho \left(\frac{\partial u}{\partial t} + u\frac{\partial u}{\partial x} + v\frac{\partial u}{\partial y} + w\frac{\partial u}{\partial z} \right) + \frac{\partial p}{\partial x} = \mu \left(\frac{\partial^2 u}{\partial x^2} + \frac{\partial^2 u}{\partial y^2} + \frac{\partial^2 u}{\partial z^2} \right)$$

$$\rho \left(\frac{\partial v}{\partial t} + u\frac{\partial v}{\partial x} + v\frac{\partial v}{\partial y} + w\frac{\partial v}{\partial z} \right) + \frac{\partial p}{\partial y} = \mu \left(\frac{\partial^2 v}{\partial x^2} + \frac{\partial^2 v}{\partial y^2} + \frac{\partial^2 v}{\partial z^2} \right)$$

$$\rho \left(\frac{\partial w}{\partial t} + u\frac{\partial w}{\partial x} + v\frac{\partial w}{\partial y} + w\frac{\partial w}{\partial z} \right) + \frac{\partial p}{\partial z} = \mu \left(\frac{\partial^2 w}{\partial x^2} + \frac{\partial^2 w}{\partial y^2} + \frac{\partial^2 w}{\partial z^2} \right)$$

$$\frac{\partial u}{\partial x} + \frac{\partial v}{\partial y} + \frac{\partial w}{\partial z} = 0$$

Comparing with Eqs.2.8.1–3, 2.4.8, it is seen where additional terms arise because of curvature in cylindrical polar coordinates. While the rectangular Cartersian form is simpler, however, it is not suitable for flow in a tube because of the cylindrical geometry of the tube.

Chapter 3

1. When flow enters a tube, only fluid in contact with the tube wall is initially influenced by the presence of the tube wall and the no-slip boundary condition there (Fig.3.1.1). As fluid moves further downstream, the region of influence grows toward the tube axis until eventually all fluid becomes affected by the presence of the tube wall. A condition is then reached whereby the flow no longer changes as it moves further downstream, that is, it becomes fully developed. Analytically, the velocity u no longer changes in x, the axial direction, hence

$$\frac{\partial u}{\partial x} \equiv 0$$

2. In addition to assumptions on which the Navier–Stokes equations and equation of continuity are based, Eq.3.2.9 for steady or pulsatile flow in a rigid tube is based on the assumption that the tube is perfectly cylindrical and its cross section is perfectly circular so that the flow field is symmetrical about the axis of the tube and the angular velocity is identically zero. It is further assumed that the tube is sufficiently long for the flow to be fully developed, hence the radial velocity is identically zero and flow is in the axial direction only.

3. The minus sign in this relation indicates that the velocity $\hat{u}_s$ and pressure gradient k_s must have opposite signs, which is as it should be on physical grounds. Flow in a tube is in the direction of *decreasing* pressure, that is, in the direction of negative pressure gradient.

4. For this purpose consider a cylindrical volume of fluid of radius r and length δx positioned along the axis of the tube. In Poiseuille flow the acceleration is zero and the fluid is in mechanical equilibrium under the viscous shear stress τ_{rx} acting on its wall and pressure force p_s acting on its two ends. That is, the sum of the two forces must be zero:

$$\left\{ p_s - \left(p_s + \frac{dp_s}{dx} \right) \right\} \pi r^2 \delta x + \tau_{rx} (2\pi r) \delta x = 0$$

Upon simplification and substitution for τ_{rx} from Eq.3.4.5, this gives

$$\frac{dp_s}{dx} = \frac{2\mu}{r} \frac{du_s}{dr}$$

This equation is Eq.3.3.2 integrated once subject to the condition of symmetry

$$\left(\frac{du_s}{dr} \right)_{r=0} = 0$$

Alternatively, the equation obtained from a balnce of forces differentiated once with respect to r and added to itself yields Eq.3.3.2.

5. Solving Eq.3.3.2 as before but now with $u_s(a) = u^*$ instead of $u_s(a) = 0$ gives

$$u_s = u^* + \frac{k_s}{4\mu}(r^2 - a^2)$$

and integrating as in Eq.3.4.3 to get the flow rate, gives

$$q_s = \pi a^2 u^* - \frac{k_s \pi a^4}{8\mu}$$

6. In the absence of slip, and with viscosity μ^*, the flow rate would be given by (Eq.3.4.3)

$$q_s = -\frac{k_s \pi a^4}{8\mu^*}$$

Equating this to the flow rate in the previous exercise gives

$$-\frac{k_s \pi a^4}{8\mu^*} = \pi a^2 u^* - \frac{k_s \pi a^4}{8\mu}$$

which yields the required result upon simplification.

7. Denoting the power by H_1 when the tube radius is a, and by H_2 when the tube radius is $0.9a$, then using Eq.3.4.14 we have

$$H_1 = \frac{8\mu l q^2}{a^4}, \quad H_2 = \frac{8\mu l q^2}{a^4(0.9)^4}$$

The required additional power is then given by

$$100 \times \frac{H_2 - H_1}{H_1} \approx 52\%$$

8. Using the notation of Section 3.7, then in the case of a more general power law Eq.3.7.4 becomes

$$a_0^n = a_1^n + a_2^n$$

while Eqs.3.7.5 become

$$\frac{a_1}{a_0} = \frac{1}{(1+\alpha^n)^{1/n}}, \quad \frac{a_2}{a_0} = \frac{\alpha}{(1+\alpha^n)^{1/n}}$$

and the two results give

$$\beta = \left(\frac{a_1}{a_0}\right)^2 + \left(\frac{a_2}{a_0}\right)^2 = \frac{1+\alpha^2}{(1+\alpha^n)^{2/n}}$$

For a symmetrical bifurcation ($\alpha = 1.0$) and $n = 2, 4$ this gives $\beta = 1.00$ and $\beta \approx 1.41$ respectively, compared with $\beta \approx 1.26$, obtained from the cube law.

9. Equations 3.4.7,14 indicate that the wall shear stress τ and pumping power H depend on the flow rate q and tube radius a as follows

$$\tau \propto \frac{q}{a^3}, \quad H \propto \frac{q^2}{a^4}$$

Thus according to the cube law ($n = 3,\ q \propto a^3$) the shear stress would be the same at all levels of the tree structure, that is $\tau = const.$, while the pumping power would be higher in vessels of larger diameter and diminish at smaller branches, more precisely $H \propto a^2$. The corresponding results for $n = 2$ ($q \propto a^2$) would be: $\tau \propto 1/a,\ H = const.$, and for $n = 4$ ($q \propto a^4$): $\tau \propto a,\ H \propto a^4$.

10. Let a_0 be the radius of the single tube and a_1 be that of each of the two tubes. The area ratio is then given by

$$\beta = \frac{2a_1^2}{a_0^2}$$

The flow rate in each of the two tubes being half of that in the single tube, the pumping powers for flow in the single tube and in each of the two tubes are then given by (Eq.3.4.14)

$$H_0 = \frac{8\mu l q^2}{\pi a_0^4}, \quad H_1 = \frac{8\mu l (q/2)^2}{\pi a_1^4}$$

The required additional fractional power is then given by

$$\triangle H = \frac{2H_1 - H_0}{H_0} = \frac{2}{\beta^2} - 1$$

This result is the same as that in Eq.3.8.1 when $\beta = 2^{1/3}$. Equation 3.8.1 is based on the cube law which yields this value of β when the bifurcation is symmetrical (Eq.3.7.8).

11. According to the cube law the flow rate in a tube is proportional to d^3, where d is tube diameter. Because average velocity in the tube is equal to flow rate divided by d^2, then average velocity in Eq.3.9.1 would be proportional to d, with the result

$$\frac{l_e}{d} \propto d^2$$

Thus the entry length expressed in terms of tube diameters diminishes rapidly at higher levels of an arterial tree where branches have successively smaller diameters.

12. Using Eqs.3.10.6,7, and if the aspect ratio of the elliptic cross section is denoted by λ, Eq.3.10.9 becomes

$$\delta^4 = 8\left(\frac{\lambda}{1+\lambda^2}\right)^3 a^4$$

For $\lambda = 1.1$ this gives $\delta^4 \approx 0.99\ a^4$, thus the flow rate would decrease by approximately only one percent and the power would increase by the same amount.

Chapter 4

1. When the pressure gradient driving the flow in the fully developed region of a rigid tube changes in time, flow in the entire region changes in response and in unison. There is no wave motion in space, there is only time oscillation of the entire body of fluid, back and forth. This situation is somewhat artificial physiologically, since a blood vessel would normally be nonrigid and oscillatory flow within it is no longer in unison. There is then wave motion in space, along the vessel, as discussed in Chapter 5. This wave motion does not exist in a rigid tube.
2. Separation of the two groups of terms in Eq.4.2.2 is only possible when the equation governing the flow is linear, which in this case is Eq.3.2.9. This equation is linear because it is restricted to the fully developed region of the flow field. Therefore, Eq.4.2.2 and Eq.4.2.3 for oscillatory flow that follows from it are only valid in the fully developed region of the flow field.
3. The first four harmonics, using the given data, are given by

$$f_1(t) = 7.5803 \cos\left(\frac{2\pi t}{T} + \frac{\pi}{180} \times (-173.9168)\right)$$

$$f_2(t) = 5.4124 \cos\left(\frac{4\pi t}{T} + \frac{\pi}{180} \times (88.9222)\right)$$

$$f_3(t) = 1.5210 \cos\left(\frac{6\pi t}{T} + \frac{\pi}{180} \times (-21.7046)\right)$$

$$f_4(t) = 0.5217 \cos\left(\frac{8\pi t}{T} + \frac{\pi}{180} \times (-33.5372)\right)$$

The first approximation to the composite wave would be $f_1(t)$, the second $f_1(t) + f_2(t)$, and so on.
4. From Eq.4.4.3, by differentiation, we find

$$\frac{\partial u_\phi}{\partial r} = \frac{dU_\phi}{dr} e^{i\omega t}$$

$$\frac{\partial^2 u_\phi}{\partial r^2} = \frac{d^2 U_\phi}{dr^2} e^{i\omega t}$$

$$\frac{\partial u_\phi}{\partial t} = i\omega U_\phi e^{i\omega t}$$

Substituting these elements into Eq.4.4.2 we obtain

$$\frac{d^2 U_\phi}{dr^2} e^{i\omega t} + \frac{1}{r}\frac{dU_\phi}{dr} e^{i\omega t} - \frac{\rho}{\mu} i\omega U_\phi e^{i\omega t} = \frac{k_s}{\mu} e^{i\omega t}$$

The exponential term cancels throughout, and after introducing Ω from Eq.4.4.5, we obtain Eq.4.4.4 as required.

5. By differentiation, we have

$$U_\phi = \frac{ik_s a^2}{\mu\Omega^2}, \quad \frac{dU_\phi}{dr} = 0, \quad \frac{d^2U_\phi}{dr^2} = 0$$

and substiuting in Eq.4.4.4 we get

$$-\frac{i\Omega^2}{a^2}\frac{ik_s a^2}{\mu\Omega^2} = \frac{k_s}{\mu}$$

in which the equality is clearly satisfied.

6. The homogeneous form of Eq.4.4.4 is

$$\frac{\partial^2 U_\phi}{\partial r^2} + \frac{1}{r}\frac{\partial U_\phi}{\partial r} - \frac{i\Omega^2}{a^2}U_\phi = 0$$

From $U_\phi = AJ_0(\zeta)$, and recalling from Eq.4.5.4 that $\zeta = \Lambda r/a$, we have by differentiation

$$\frac{dU_\phi}{dr} = \frac{dU_\phi}{d\zeta}\frac{d\zeta}{dr} = A\frac{\Lambda}{a}\frac{dJ_0}{d\zeta}$$

$$\frac{d^2U_\phi}{dr^2} = A\frac{\Lambda^2}{a^2}\frac{d^2J_0}{d\zeta^2}$$

Substituting into the homogeneous equation above we get

$$A\frac{\Lambda^2}{a^2}\frac{d^2J_0}{d\zeta^2} + \frac{1}{r}A\frac{\Lambda}{a}\frac{dJ_0}{d\zeta} - \frac{i\Omega^2}{a^2}AJ_0 = 0$$

Upon multiplication by $a^2/A\Lambda^2$ throughout the equation, and using the relations between Ω, Λ, and ζ, the equation reduces to Eq.4.5.2. The same procedure with $U_\phi = BY_0(\zeta)$ leads to Eq.4.5.3.

7. For $\Omega = 3.0$ we find from Appendix A:

$$\Lambda = -2.1213 + 2.1213i$$
$$J_0(\Lambda) = -0.2214 + 1.9376i$$

At the tube center $r = 0$ and from Eq.4.5.4 $\zeta = 0$. From Appendix A we find $J_0(0) = 1$. Using these values in Eq.4.6.2, we then have, after simplifying,

$$\frac{u_\phi(0,t)}{\hat{u}_s} = (0.2264 - 0.4703i)(\cos\omega t + i\sin\omega t)$$

The velocity corresponding to the given pressure gradient is given by the *real part* of the above expression, since $k_s \cos\omega t$ is the real part of the pressure gradient. Therefore

a. At the begining of the cycle where $\omega t = 0$, the required velocity is 0.2264.

b. At a quarter way through the cycle where $\omega t = 90°$, the required velocity is 0.4703.

These two values correspond with the peak velocities in the top first and second panels of Fig.4.6.1, respectively.

8. For $\Omega = 3.0$ we find from Appendix A:

$$\begin{aligned} \Lambda &= -2.1213 + 2.1213i \\ J_0(\Lambda) &= -0.2214 + 1.9376i \\ J_1(\Lambda) &= -1.7326 - 0.4875i \end{aligned}$$

Using these values and $\omega t = 150\pi/180$ in Eq.4.7.7, we find, after simplifying,

$$\frac{q_\phi(t)}{q_s} = (0.3198 - 0.4454i)(\cos 150\pi/180 + i \sin 150\pi/180)$$

The required flow rate is the imaginary part of this, namely, 0.5456, which agrees with the peak flow in Fig.4.7.1.

9. From Eqs.4.9.16,17, with Eq.3.4.14, the oscillatory part of the pumping power is given by

$$\frac{H_{\phi R}}{H_s} = \left(\frac{k_{\phi R}}{k_s}\right)\left(\frac{q_{\phi R}}{q_s}\right) = \cos\omega t \left(\frac{q_{\phi R}}{q_s}\right)$$

$$\frac{H_{\phi I}}{H_s} = \left(\frac{k_{\phi I}}{k_s}\right)\left(\frac{q_{\phi I}}{q_s}\right) = \sin\omega t \left(\frac{q_{\phi I}}{q_s}\right)$$

Using results from the previous example, this gives

$$\frac{H_{\phi R}}{H_s} = \cos\omega t(0.3198 \cos\omega t + 0.4454 \sin\omega t)$$

$$\frac{H_{\phi I}}{H_s} = \sin\omega t(0.3198 \sin\omega t - 0.4454 \cos\omega t)$$

The average over one cycle is given by

$$\frac{1}{2\pi/\omega}\int_0^{2\pi/\omega} \frac{H_{\phi R}}{H_s} dt = \frac{1}{2\pi/\omega}\int_0^{2\pi/\omega} \frac{H_{\phi I}}{H_s} dt \approx 0.16$$

The result is in visual agreement with that in Fig.4.9.1

10. From Eq.4.10.14 for low frequency, with $\Omega = 1$, we have

$$\frac{q_{\phi R}(t)}{q_s} \approx \cos\omega t$$

while from Eq.4.11.14 for high frequency, with $\Omega = 10$, we have

$$\frac{q_{\phi R}(t)}{q_s} \approx \frac{8}{100} \sin\omega t$$

The first has a maximum of 1.0, while the second has a maximum of about 0.1, which agree visually with those in Fig.4.11.3. The same results are obtained by using the imaginary parts of the flow rate.

11. From Eqs.4.6.3,4 the real part of the pressure gradient varies as $\cos \omega t$ while the imaginary part varies as $\sin \omega t$. Thus the real part has its peak at $\omega t = 0$, while the imaginary part has its peak at $\omega t = 90°$.
From results of the previous example we see that at low frequency ($\omega = 1$) the real part of the flow rate varies approximately as $\cos \omega t$, while at high frequency ($\omega = 10$) it varies as $(8/100) \sin \omega t$. The peak of the first expression occurs at $\omega t \approx 0$, while that of the second occurs at $\omega t = 90°$. Thus at low frequency the flow rate is almost in phase with the pressure gradient, while at high frequency it lags almost 90° behind it. These values agree visually with those in the lower panel of Fig.4.11.3. The same results are obtained by considering the imaginary part of the flow rate.

12. From Eq.4.10.24 and Eq.3.4.14 we have

$$\frac{H_{\phi R}}{H_s} = \left(\frac{k_{\phi R}}{k_s}\right)\left(\frac{q_{\phi R}}{q_s}\right)$$

For low frequency, using Eq.4.10.14 and Eqs.4.6.3,4 this gives

$$\frac{H_{\phi R}}{H_s} \approx \cos^2 \omega t$$

The average of this quantity over one cycle, that is, from $\omega t = 0$ to $\omega t = 2\pi$, is 1/2. Thus at low frequency the wasted power in oscillatory flow is about one half of the corresponding pumping power in steady flow. The same results are obtained by considering the imaginary part of the flow rate.
Similarly, at high frequency using Eqs.4.11.18,19 we find

$$\frac{H_{\phi R}}{H_s} \approx \frac{8}{\Omega^2} \cos \omega t \sin \omega t$$

This quantity has zero average over one cycle, that is, from $\omega t = 0$ to $\omega t = 2\pi$. Thus at high frequency the wasted power in oscillatory flow is close to zero.

Chapter 5

1. The two additional assumptions on which the equations governing pulsatile flow in an elastic tube are based are (a) that the propagating wavelength is much larger than the tube radius and (b) that the propagating wave speed is much higher than the average flow velocity within the tube.

2. A stiffer tube has a higher modulus of elasticity E and therefore, from Eq.5.1.1, higher wave speed c_0. Thus, all else being equal, the first assumption (see previous exercise) is better satisfied in a stiffer tube. In a tube of smaller radius, all else being equal, the length of the propagating wave is more likely to be larger than the tube radius, hence the second assumption (see previous exercise) would be better satisfied in a tube of smaller radius.

3. It may be argued that this question is not a legitimate one because in the case of pulsatile flow in a rigid tube *there is no wave motion.* It is more instructive, however, to view a rigid tube as a limiting case of an elastic tube that is infinitely stiff (see previous exercise), therefore having an infinite modulus of elasticity E. Equation 5.1.1 would then indicate that the wave speed c_0 is infinite. Thus pulsatile flow in a rigid tube may be viewed as a limiting case of pulsatile flow with infinite wave speed.

4. Equations 5.2.4–6 governing motion of the fluid in an elastic tube are *linear* equations in the dependent variables U, V, P, which makes them amenable to a standard solution. If the boundary conditions are applied at a *moving* boundary, a standard solution is no longer possible, in fact the solution becomes almost intractable mathematically. Application of the boundary conditions at a constant "neutral" radius a is motivated in the first instance by this difficulty but can also be justified on physical grounds if it can be assumed that radial movements of the tube wall are small. In the physiological setting this assumption is reasonably well satisfied because of the confinement of blood vessels by surrounding tissue and because the modulus of elasticity of the wall is fairly high.

5. There are four forces acting on a segment of the tube wall:

a. Axial tension within the tube wall, resulting from nonuniform axial pull at different axial positions along the tube wall.
b. Radial force within the tube wall, resulting from *circumferential* tension within the wall.
c. Radial force acting on the tube wall by fluid pressure within the tube.
d. Axial force acting on the tube wall by the moving fluid and resulting from viscous shear at the tube wall.

6. Equations 5.4.16,17 are statements of Newton's law of motion in the axial and radial directions, respectively, applied to a segment of the tube wall. In the first equation, axial acceleration on the left-hand side equals the net of shear force and axial tension within the tube wall on the right-hand side, divided by the mass of the tube segment. In the second equation, radial acceleration on the left-hand side equals the net of pressure force and radial stress within the tube wall on the right-hand side, again divided by the mass of the tube segment.

7. The assumption that the tube wall thickness is much smaller than the tube radius arises at two main points in the analysis. First, when considering the radial stress component within the tube wall (Eq.5.3.4), a force resulting from a stress gradient *within the thickness of the tube wall* is neglected on the assumption that the wall is thin. Second, the stress–

strain relations (Eqs.5.4.6,7) are much simplified on the assumption that the tube wall is thin.

8. The coupling between the equations governing fluid and wall movements occurs because Eqs.5.4.16,17 for wall movement contain elements of the flow field, namely, pressure p_w and shear stress τ_w at the tube wall. Therefore, the two sets of equations cannot be solved independently of each other, they must be solved *simultaneously.*

9. Because the two sets of equations, governing fluid flow and wall movement, must be solved simultaneously, some boundary conditions are shared between the two sets. These are then referred to as "matching conditions" because they arise from matching of physical movements of the fluid and of the tube wall. Because the two movements come together at the interface between the fluid and the tube wall, this is where the matching conditions are applied (Eqs.5.6.1,2).

10. The wave speed c_0 defined by Eq.5.1.1 does not take into account the viscous coupling between fluid and wall movements. Hence it is referred to as the *inviscid* wave speed, since the viscous shear force exerted by the moving fluid is being neglected, as if the fluid were inviscid. The *actual* wave speed c in an elastic tube, as defined by Eq.5.7.21, does take the viscous effect into account since it is obtained by solving the coupled equations. In fact, as may be expected on physical grounds, viscosity reduces the speed with which the wave propagates in an elastic tube, as can be observed from Fig.5.7.1.

11. For $\Omega = 1$, we read from Appendix A:

$$\Lambda = -0.7071 + 0.7071i$$
$$J_0(\Lambda) = 0.9844 + 0.2496i$$
$$J_1(\Lambda) = -0.3959 + 0.3076i$$
$$G(\Lambda) = 1.0049 - 0.0014i$$

Using Eqs.4.7.7, 5.7.20, 5.9.9, we find

$$g = 0.9798 - 0.1216i$$
$$\left|\frac{q_\phi}{q_s}\right| \text{(rigid)} = 0.9860$$
$$\left|\frac{q_\phi}{q_s}\right| \text{(elastic)} = 0.9963$$
$$\text{difference} \approx 1\%$$

Repeating for $\Omega = 3$ we find

$$g = 0.4990 - 0.3599i$$
$$\left|\frac{q_\phi}{q_s}\right| \text{(rigid)} = 0.5484$$
$$\left|\frac{q_\phi}{q_s}\right| \text{(elastic)} = 0.6252$$

$$\text{difference} \approx 14\%$$

and for $\Omega = 10$ we find

$$g = 0.1416 - 0.1312i$$

$$\left|\frac{q_\phi}{q_s}\right| (\text{rigid}) = 0.0695$$

$$\left|\frac{q_\phi}{q_s}\right| (\text{elastic}) = 0.0721$$

$$\text{difference} \approx 4\%$$

The results agree visually with those in Fig.5.9.9.

Chapter 6

1. Wave reflections in a tube require the presence of a propagating wave along the tube. In the case of a rigid tube there is no wave propagation along the tube, as was seen in Chapter 4, therefore there is no wave reflections under the conditions studied in that chapter. A way in which wave propagation and wave reflections *can* occur in a rigid tube is when the fluid within the tube is compressible. In that case compressibility of the fluid takes the place of elasticity of the tube wall and the phenomenon unfolds in essentially the same way.
2. The ultimate purpose of pulsatile flow analysis is to determine the relationship between flow and pressure gradient, as was done for steady flow in Chapter 3. The same was done for pulsatile flow in Chapters 4 and 5, but the results there are only valid in the absence of wave reflections. When wave reflections are present, the relation between flow and pressure gradient can no longer be obtained from a single solution of the governing equations. All sources of wave reflections may have an effect on that relation and must be considered along with a solution of the governing equations. From the standpoint of physics, and stated simply, the flow in a tube is then no longer determined by only the driving pressure gradient.
3. While the final form of the one-dimensional wave equations (Eqs.6.2.18, 19) does not contain the radial coordinate r, the equations contain the wave speed c_0, which depends on the cross sectional area of the tube (Eq.6.2.14) and hence on the tube radius. Furthermore, the wave equation for the pressure contains the time rate of change of pressure, which in turn depends on the time rate of change of cross-sectional area and hence of tube radius. Thus while the radial coordinate r has been eliminated, the tube radius and its time rate of change remain implicitly in the equations.

4. From Eq.6.3.6, by differentiation, we have

$$\frac{dp_x}{dx} = -\frac{i\omega}{c_0}Be^{-i\omega x/c_0} + \frac{i\omega}{c_0}Ce^{i\omega x/c_0}$$

$$\frac{d^2p_x}{dx^2} = -\frac{\omega^2}{c_0^2}Be^{-i\omega x/c_0} - \frac{\omega^2}{c_0^2}Ce^{i\omega x/c_0}$$

Substituting in Eq.6.3.5 we have

$$\begin{aligned}\frac{d^2p_x}{dx^2} + \frac{\omega^2}{c_0^2}p_x(x) &= -\frac{\omega^2}{c_0^2}(Be^{-i\omega x/c_0} + Ce^{i\omega x/c_0}) \\ &\quad +\frac{\omega^2}{c_0^2}(Be^{-i\omega x/c_0} + Ce^{i\omega x/c_0}) \\ &= 0\end{aligned}$$

thus the equation is satisfied. This is the *general* solution since the equation is second order and the solution contains two arbitrary constants.

5. From Eq.6.3.4, by differentiation, we have

$$\frac{\partial p}{\partial t} = p_x i\omega e^{i\omega t}$$

$$\frac{\partial^2 p}{\partial t^2} = -p_x\omega^2 e^{i\omega t}$$

$$\frac{\partial p}{\partial x} = \frac{dp_x}{dx}e^{i\omega t}$$

$$\frac{\partial^2 p}{\partial x^2} = \frac{d^2p_x}{dx^2}e^{i\omega t}$$

Substituting in Eq.6.2.18 gives

$$\frac{\partial^2 p}{\partial t^2} - c_0^2\frac{\partial^2 p}{\partial x^2} = -p_x\omega^2 e^{i\omega t} - c_0^2\frac{d^2p_x}{dx^2}e^{i\omega t} = 0$$

which reproduces Eq.6.3.5 upon simplification.

6. From Eqs.6.4.10,17 we have

$$\begin{aligned}\overline{p}(x,t) &= \overline{p}(x)e^{i\omega t} \\ &= ((R+1)\cos 2\pi\overline{x} + i(R-1)\sin 2\pi\overline{x}) \times \\ &\quad (\cos\omega t + i\sin\omega t)\end{aligned}$$

We note in Fig.6.4.1 that individual curves represent the *real* part of the pressure distribution and that wave reflections are at 80%. Thus taking the real part of the above with $R = 0.8$ and $\omega t = \pi/2$, we find

$$\Re\{\overline{p}(x)\} = 0.2\sin 2\pi\overline{x}$$

This curve can be identified in Fig.6.4.1 as that having a value of 0 at $\overline{x} = 0$, 0.2 at $\overline{x} = 0.25$, -0.2 at $\overline{x} = 0.75$, and 0 again at $\overline{x} = 1.0$.

7. In the absence of wave reflections $R = 0$ and Eq.6.4.18 reduces to $|\overline{p}_x(\overline{x})| = 1$, which is the equation of the outer envelope in Fig.6.3.1. When $R = 0.8$, Eq.6.4.18 becomes

$$|\overline{p}_x(\overline{x})| = \sqrt{1.64 + 1.6\cos 4\pi\overline{x}}$$

which gives $|\overline{p}_x(\overline{x})| = 1.8$ at $\overline{x} = 0$, $|\overline{p}_x(\overline{x})| = 0.2$ at $\overline{x} = 1/4$, and $|\overline{p}_x(\overline{x})| = 1.8$ again at $\overline{x} = 1/2$, thus identifying the shape of the outer envelope in Fig.6.4.1.

8. Using Eqs.6.5.19 and 6.4,12

$$\begin{aligned}
\epsilon &= R_0 R_l e^{-i\omega 2l/c_0} = R_0 R_l e^{-i4\pi l/L} \\
&= R_0 R_l e^{-i4\pi} \quad \text{when } l = L \\
&= R_0 R_l \cos 4\pi = R_0 R_l = 0.25
\end{aligned}$$

9. Using Eq.6.6.15 and the given data, we have for the tube of $1\,cm$ radius

$$\begin{aligned}
Y_0 &= \frac{A}{\rho c_0} = \frac{\pi \times 1^2 (cm^2)}{1(g/cm^3) \times 100(cm/s)} \\
&= \frac{\pi}{10^2}\left(\frac{cm^3/s}{dynes/cm^2}\right)
\end{aligned}$$

The physical dimensions of the final result are seen to be that of flow rate (cm^3/s) over pressure $(dynes/cm^2)$. The corresponding result for a tube of $1mm$ radius would be $\pi/10^4$.

10. From Eqs.6.7.1 and 6.4.12, we have

$$\begin{aligned}
\frac{Y_e}{Y_0} &= \frac{1 - Re^{-2i\omega l/c_0}}{1 + Re^{-2i\omega l/c_0}} \\
&= \frac{1 - Re^{-4\pi i l/L}}{1 + Re^{-4\pi i l/L}} \\
&= \frac{1 - R\cos 4\pi}{1 + R\cos 4\pi} \quad \text{when } l = L \\
&= \frac{1 - R}{1 + R} = \frac{1}{3} \quad \text{when } R = 0.5
\end{aligned}$$

11. Because characteristic admittance is proportional to the cross-sectional area of a tube (Eq.6.6.15), Eq.6.7.14 applied to the arterial bifurcation gives

$$\begin{aligned}
R &= \frac{a_0^2 - (a_1^2 + a_2^2)}{a_0^2 + (a_1^2 + a_2^2)} \\
&= \frac{1 - \beta}{1 + \beta}
\end{aligned}$$

where a_0, a_1, a_2 are radii of the parent vessel and two branches, respectively, and β is the area ratio as defined in Eq.3.7.2. If the bifurcation is symmetrical and the cube law is in effect, $\beta = 2^{1/3}$ (Eq.3.7.8), which gives $R \approx -0.115$.

12. The passage from one level to the next in a tree structure is characterized by the bifurcation of each tube segment into two branches. It is therefore sufficient for the purpose of this question to consider a single bifurcation in which the radius of the parent tube is a_0 and those of the two branches are a_1, a_2. If the corresponding *characteristic* admittances are denoted by Y_0, Y_1, Y_2, and if we recall from Eq.6.6.15 that characteristic admittance is proportional to cross-sectional area and hence to the square of tube radius, then

$$\frac{Y_1 + Y_2}{Y_0} = \frac{a_1^2 + a_2^2}{a_0^2} = \beta$$

From this we see that the ratio of combined admittance of the two branches to that of the parent tube is equal to the area ratio β. Since the area ratio generally increases from one level to the next in a vascular tree structure (moving from parents to branches), the characteristic admittance would also increase in that direction.

If wave reflections are to be taken into account, we must consider whether the combined *effective* admittance of the two branches would be higher or lower than their combined *characteristic* admittance. The answer to this depends on the hierarchic structure of the tree upstream of this particular bifurcation. The example cited in Section 6.8 indicates that the effective admittances of the branches are higher than their characteristic admittances in that particular tree structure.

Index

Continued from page ii

PUBLISHED VOLUMES:

Russell K. Hobbie: Intermediate Physics for Medicine and Biology, *Third Edition*

George J. Hademenos and Tarik F. Massoud: The Physics of Cerebrovascular Diseases

M. Zamir: The Physics of Pulsatile Flow

IN PREPARATION:

George B. Benedek and Felix M.H. Villars: Physics with Illustrative Examples from Medicine and Biology, Mechanics, *Second Edition*

George B. Benedek and Felix M.H. Villars: Physics with Illustrative Examples from Medicine and Biology, Statistical Physics, *Second Edition*

George B. Benedek and Felix M.H. Villars: Physics with Illustrative Examples from Medicine and Biology, Electricity and Magnetism, *Second Edition*